Ecological Bulletins No. 50

BIODIVERSITY EVALUATION TOOLS FOR EUROPEAN FORESTS

Tor-Björn Larsson, Per Angelstam, Gérard Balent, Anna Barbati, Rienk-Jan Bijlsma,
Andrej Boncina, Richard Bradshaw, Winfried Bücking, Orazio Ciancio, Piermaria Corona,
Jurij Diaci, Susana Dias, Herrmann Ellenberg, Francisco Manuel Fernandes,
Frederico Fernández-Gonzalez, Richard Ferris, Georg Frank, Peter Friis Møller, Paul S. Giller,
Lena Gustafsson, Klaus Halbritter, Susannah Hall, Lennart Hansson, John Innes, Hervé Jactel,
Michéle Keannel Dobbertin, Manfred Klein, Marco Marchetti, Frits Mohren, Pekka Niemelä,
John O'Halloran, Ewald Rametsteiner, Francisco Rego, Christoph Scheidegger, Roberto Scotti,
Kjell Sjöberg, Ionnis Spanos, Konstantinos Spanos, Tibor Standovár, Linus Svensson,
Bjørn Åge Tømmerås, Dimitris Trakolis, Janne Uuttera, Diego VanDenMeersschaut,
Kris Vandekerkhove, Paul M. Walsh, Allan D. Watt

"The study has been carried out with financial support from the Commission of the European Communities, Agriculture and Fisheries (FAIR) specific RTD programme, CT-3575 "Indicators for monitoring and evaluation of forest biodiversity in Europe". It does not necessarily reflect its views and in no way anticipates the Commission´s future policy in this area".

BIODIVERSITY EVALUATION TOOLS FOR EUROPEAN FORESTS

Contents

How to use this report 7

Executive summary 7

Chapter 1. The biodiversity challenge 11
    1.1 Biodiversity and its components 11
    1.2 Policy development as regards the management of forest biodiversity 13
    1.3 Biodiversity assessment and evaluation 15
    1.4 The EU project "Indicators for monitoring and evaluation of forest biodiversity in Europe, BEAR" 15

Chapter 2. Biodiversity in European forests 19
    2.1 Historical development of the forest land 19
    2.2 The state of biodiversity in European forests 21

Chapter 3. Identifying main features of European forest biodiversity 27
    3.1 Key factors of forest biodiversity 27
    3.2 The concept of forest types for biodiversity assessment 34

Chapter 4. Major European forest types for biodiversity assessment 39

Chapter 5. Principles for assessing forest biodiversity in Europe 127
    5.1 Biodiversity indicators and assessment 127
    5.2 Assessing the key factors of forest biodiversity 129

Chapter 6. Conclusions 141
    6.1 A strategy for assessment of forest biodiversity on European scalel 141
    6.2 Assessment of forest biodiversity on the operational scale 142
    6.3 The need for further development and research 143

References 145

Appendix 1. BEAR partners and user-panel 151
Appendix 2. Important European forest types according to the habitats directive 159
Appendix 3. CORINE land cover types of relevance to forest biodiversity 161
Appendix 4. The scheme of potential natural vegetation of Europe 163
Appendix 5. European forest type schemes 167
Appendix 6. A bibliography of terms and definitions 171
Appendix 7. Indicators of biodiversity: recent approaches and some general suggestions 223
Appendix 8. Electronic conference on research and biodiversity: preliminary report on the session on forest 231

Ecological Bulletins

ECOLOGICAL BULLETINS are published in cooperation with the ecological journals Ecography and Oikos. Ecological Bulletins consists of monographs, reports and symposia proceedings on topics of international interest, often with an applied aspect, published on a non-profit making basis. Orders for volumes should be placed with the publisher. Discounts are available for standing orders.

Editor-in-Chief and Editorial Office:
Pehr H. Enckell
Oikos Editorial Office
Ecology Building
SE-223 62 Lund
Sweden
Fax: +46-46 222 37 90
Email: oikos@ekol.lu.se
www.oikos.ekol.lu.se

Technical Editors:
Linus Svensson/Gunilla Andersson

Editorial Board:
Björn E. Berglund, Lund.
Tom Fenchel, Helsingør.
Erkki Leppäkoski, Turku.
Ulrik Lohm, Linköping.
Nils Malmer (Chairman), Lund.
Hans M. Seip, Oslo.

Published and distributed by:
Blackwell Science
Sales Department
108 Cowley Road
Oxford
U.K. OX4 1JF
Tel: +44-1865 791 100
Fax: +44-1865 791 347
Email: direct.order@marston.co.uk
www.blackwell-science.com

Suggested citation:
Larsson, T.-B. 2001. Biodiversity Evaluation Tools for European forests. – Ecol. Bull. 50: 000–000.

This volume of ECOLOGICAL BULLETINS can be ordered at
www.oikos.ekol.lu.se

© 2001, ECOLOGICAL BULLETINS
ISBN 87-16-16434-2
ISSN 0346-6868

Cover: Elias Martin (1739–1818): Bergigt landskap [Mountainous landscape], Univ. Library, Uppsala, Sweden.

Typeset by ZooBoTech/GrafikGruppen, Torna Hällestad, Sweden, printed by Wallin & Dalholm, Lund, Sweden.

How to use this report

The objective of this report is to give recommendations for the development of Biodiversity Evaluation Tools (BETs) for European forests adapted to different specific use, e.g. the following:

- Provide forest managers with an operational tool for stand level discrimination of types of forest under different structural, functional, management conditions with due regard to biodiversity conservation issues;
- Supply sub-national (regional/local) administrational forestry organisations with an evaluation tool to assess, at the landscape scale, the effects of local forest management on the maintenance of biodiversity and to feed back results into the planning process;
- Help the national organisations responsible for reporting for EU habitat conservation strategies to assess the quality of Sustainable Forest Management strategies in terms of biodiversity conservation;
- Provide the Ministerial Process for Protection of Forest in Europe with a basis for improving the implementation of the Helsinki Resolution H2 "General guidelines for the Conservation of Biodiversity in European forests".

We neither have the knowledge nor the resources to perform a complete assessment of European forest biodiversity. This is no excuse for not taking action to mitigate the great threats to the biodiversity of European forests. The forestry sector has a tradition to pragmatically address silvicultural issues and the uncertainty as regards knowledge should not to be an obstacle to action.

A positive fact is the intense ongoing research and development activities with respect to biodiversity assessment both in forests and other ecosystems, in national, European and other international programmes. A strong recommendation is to follow the current research and development projects, on an international scale, and be prepared to take on board relevant innovations.

Concerning this report it must be stated that we do not recommend any organisation to directly put all discussed indicators into operational work. The "finalised knowledge" and precise instructions on assessment of biological diversity in practical management must be the responsibility of managing organisation and/or the policy-makers deciding on the legislative and economic instruments; including the negotiations to take place in the process for protection of forests in Europe. Please read the following chapters as a suggested work model and a list of options for assessing the biodiversity of European forests – and good luck in your work!

Executive summary

The project "Indicators for monitoring and evaluation of forest biodiversity in Europe BEAR", initiated in 1998, is a pan-European concerted action, bringing together expertise from 27 European research organisations to build a framework for the development of forest biodiversity indicators at various scales. The six major biogeographic regions of Europe are thus represented by partners from the following countries (cf. Table 2, Appendix 1):

- Boreal region: Sweden, Finland, Norway;
- Continental region: Germany, Hungary, Austria, Switzerland, Italy;
- Atlantic region: U.K., Ireland, Belgium, France, Denmark, the Netherlands;
- Alpine region: Switzerland, Austria, Italy, Slovenia;
- Macaronesian region: Madeira, Portugal, Spain;
- Mediterranean region: Portugal, Spain, Greece, Italy, France.

The work has been undertaken over a two years-period, to meet the following specific aims:

1. To analyse the important forest types in the six major European biogeographic regions with respect to their structure and function, in order to identify key parameters and determinants of biodiversity, adapted to the national level, the landscape level and the stand level;
2. To harmonise the classification of important forest types and suggest forest biodiversity indicators to assess key factors of biodiversity;
3. To summarise the applicability of indicators as regards ecological significance and data availability, including comments on assessment methodologies;
4. To discuss the recommendations with end-users, with particular reference to indicator feasibility;
5. To synthesise the results into a strategic manual "Biodiversity Evaluation Tools (BETs)", which can be defined as the combination of biodiversity indicators and the standardised methodology needed to apply them.

The main achievements of BEAR are:

1. AGREEMENT ON A COMMON SCHEME OF KEY FACTORS OF BIODIVERSITY APPLICABLE TO EUROPEAN FORESTS. "Key factors" affecting or determining biodiversity include abiotic, biotic and anthropogenic factors that directly or indirectly influence biodiversity and its major components (composition, stucture and function). The BEAR experts have agreed on a single common scheme of key factors relevant to all European forests – see Table 4.
2. IDENTIFYING EUROPEAN-LEVEL FOREST TYPES FOR BIODIVERSITY ASSESSMENT FTBAs. The relative importance of key factors vary between different European forests as do the factors themselves, e.g. the species composition. The BEAR experts recommend that the management of bio-

diversity is based upon specific Forest Types for Biodiversity Assessment FTBAs. Each FTBA is distinguished by its particular features, i.e. key factors of forest biodiversity. A preliminary list of ca 30 FTBAs to be taken into account on a European level is suggested, see Table 6. A description of each FTBA is presented in Chapter 4. Each FTBA has been defined to correspond to one or several units of a European Map of Potential Natural Vegetation (scale of 1:10 million). Furthermore, it is possible to relate the FTBAs to the CORINE classifications as well as to the priority habitats listed in the Habitats Directive.

3. INDICATORS OF FOREST BIODIVERSITY. The biodiversity indicators are the measures used to assess the key factors of forest biodiversity. For operational reasons the indicators need to be adapted to the special conditions of each Forest Type for Biodiversity Assessment (FTBA) as well as to the scale and other prerequisites of the specific use. During the BEAR project the experts have agreed on that it is premature to define priority lists of indicators for operational use. This view has been proposed and accepted as the current EU position, with reference to the Convention on Biological Diversity. However, the BEAR project has presented a list of potential biodiversity indicators to assess each key factor of forest biodiversity, cf. Chapter 5.

4. RECOMMENDATIONS FOR ELABORATING "BIODIVERSITY EVALUATION TOOLS (BETs)" AND ESTABLISHING SCHEMES OF BIODIVERSITY INDICATORS FOR ASSESSMENT OF FOREST BIODIVERSITY ON A EUROPEAN SCALE. The BEAR project gives the following general advice, in priority order, for application in European and national scale assessment and monitoring of forest biodiversity (more detailed recommendations are given in Chapter 6).

INTRODUCE THE KEY FACTOR APPROACH IN EUROPEAN AND NATIONAL SCALE MONITORING OF FOREST BIODIVERSITY: A summary of the state and trends in a set of key factors, as presented in the BEAR project, should be the most efficient measure of the state and trends in European forest biodiversity. A priority list of key factors should be agreed on, but indicators to assess these key factors must not necessarily be standardised, as a further development of indicator schemes and methodology is desirable.

MAKE A FURTHER DIVISION INTO FOREST TYPES FOR BIODIVERSITY ASSESSMENT (FTBAs) IN THE REPORTING OF EACH KEY FACTOR: A great improvement in the reporting of state and trends of forest biodiversity can be brought about through the assessment of key factors with respect to the Forest Types for Biodiversity Assessment (FTBAs). Also assesing the ACTUAL total area of each FTBA/country is a high priority task. Compatibility with other systems, e.g. CORINE, should be ensured, as long as the basic principles of delineating the FTBAs are not violated. In the long term, a more nature-based approach to European-scale

biodiversity management would be to abandon the countries as units in favour of biogeographic regions.

The BEAR project advocates giving priority to the first two recommendations, above, but taking a long-term perspective:

TOWARDS STANDARDISED INDICATORS, METHODOLOGY AND PROTOCOLS: Taking a long-term perspective, it is necessary for a standardised system of indicators of forest biodiversity to be implemented in a global assessment of state and trends in biological diversity. Europe may, with the aid of strong co-ordination by the European Commission, take the lead this work. The BEAR project advocates further work to develop standardised indicators along two lines, but in combination to optimise cost-effectiveness.
1. Expand the National Forest Inventories (NFIs) to encompass indicators of forest biodiversity;
2. Introduce satellite-based techniques for European-scale monitoring of forest biodiversity.

5. RECOMMENDATIONS FOR ELABORATING "BIODIVERSITY EVALUATION TOOLS (BETs)" AND ESTABLISHING SCHEMES OF BIODIVERSITY INDICATORS FOR ASSESSMENT OF FOREST BIODIVERSITY ON THE OPERATIONAL UNIT LEVEL. Planning for biodiversity on the operational unit level may occur at several spatial levels. From an operational forestry point of view it is adequate to discuss three scales: Single tree, stand and landscape. One may distinguish three categories of "forest managers" that could be assigned different major "responsibilities": small-scale forestry, large-scale forest enterprises and the regional/national/ governmental. Brief recommendations are given for elaboration of BETs for each scale and forest manager category.

6. HIGHLIGHTING THE STATE OF THE ART OF KNOWLEDGE TO PRESENT BIODIVERSITY INDICATORS AND THE NEED FOR FUTURE RESEARCH. It is demonstrated that because of the strong development of biodiversity indicators and the insufficient current knowledge it is premature to establish a standard monitoring scheme based on "core" indicators. Five priority research areas to further develop Biodiversity Evaluation Tools for European forests and introduce these in operational use are identified:
   • VALIDATION OF INDICATORS OF FOREST BIODIVERSITY: The predictive value of selected indicators as regards (other) components of biodiversity needs to be investigated.
   • ESTABLISHMENT OF REFERENCE VALUES AND CRITICAL THRESHOLDS FOR FOREST BIODIVERSITY: A reconstruction of reference situations of forest biodiversity (e.g. near-native state, pre-industrial state) expressed through indicators values would be of practical use for setting biodiversity targets. This also applies for critical thresholds for biodiversity loss in deviation from reference situations.

• BIODIVERSITY EVALUATION TOOLS FOR ASSESSMENT OF GENETIC DIVERSITY OF FOREST TREES: The BEAR project has not taken genetic aspects into account, but research into this field is badly needed.
• DEVELOPMENT OF INDICATORS FOR NATIONAL MONITORING OF FOREST BIODIVERSITY: The National Forest Inventories should be expanded to encompass biodiversity measurements.

• STRENGTHENING LANDSCAPE-SCALE PLANNING FOR FOREST BIODIVERSITY: In particular participatory planning models for small-scale forestry need to be developed.

The BEAR project has, hopefully, contributed to an improvement in ecological understanding, communication and awareness among end-users responsible for the policy framework and operational management of European forest biodiversity.

An ecosystem has three major components: structure (e. g. physical aspects), composition (e. g. species) and function (e. g. regulating mechanisms). Forest biodiversity should be analysed accordingly.

ECOSYSTEM

The relative importance of the key factors vary among forests. A scheme of European Forest Types For Biodiversity Assessment has thus been elaborated.

BIODIVERSITY ASSESSMENT

A forest ecosystem can be characterised by key factors of biodiversity. These can be grouped according to the major ecosystem components and furthermore the scale (national/regional, landscape and stand levels) must be taken into account. A common scheme of key factors for European Forests have been elaborated.

KEY FACTORS

Indicators are the tools to assess the key factors of forest biodiversity; indicator schemes should be adapted to the specific objectives of biodiversity assessment as well as to forest types concerned.

INDICATORS

Summary of BEAR project development logic (from BEAR Newsletter no. 2, 1999).

Chapter 1.

The biodiversity challenge

1.1 Biodiversity and its components

Forests cover some 3500 million ha or 27% of earth's land surface and harbour a great proportion of all species currently known or presumed to exist (Kouki and Niemelä 1997, Bunnell and McLeod 1998, EU Commission 1998). Whilst the forests provide an important resource for humans, they are a key repository of biological diversity; and the species, communities and ecosystems they form play a central role in the functioning of the biosphere.

In 1992, Article 2 of the Convention on Biological Diversity (CBD) defined "Biodiversity" as "the variability among living organisms from all sources including, inter alia, terrestrial, marine and other aquatic ecosystems, and the ecological complexes of which they are part; this includes diversity within species, between species and of ecosystems". The term encompasses the whole range of the genetic diversity within species, the diversity of species and higher taxa, up to ecosystem diversity, and even the diversity of ecological interactions (Duelli 1997). For a review of "biodiversity" definitions see Kaennel-Dobbertin (Appendix 6).

Three primary aspects of biodiversity have been widely recognised (Franklin 1988, Noss 1990, Perry 1994, Spies 1997): composition, structure and function, cf. Table 1.

These components have been used as a basis for identification of key factors (Chapter 3.1), prior to the development of biodiversity indicators and Biodiversity Evaluation Tools (BETs).

Composition and structure determine and constitute the biodiversity of an area (Noss 1990), and are essential to the productivity and of forest ecosystems sustainability. In the present context, the "sustainable use of forests" is defined as "use that indefinitely maintains the forest both in the environmental services which it provides, as well as in its biological quality" (e.g. in a natural forest, harvest must neither exceed the regeneration rate of the resource nor impair the potential of similar harvests in the future).

Functional diversity, i.e. the diversity of ecological functions performed by different species, and/or the diversity

of species performing a given ecological function (Appendix 6), may be seen as contributing to ecological integrity; i.e. the ability of an ecosystem to function and maintain itself (Angemayer and Karr 1994, Gaston 1998, Hansson 1998) however, and its elements in turn help to generate and maintain biodiversity.

Many species are important to the continued productivity of the forest, and contribute towards ecosystem functioning. These include such diverse organisms as mycorrhizal fungi and the small mammals (Blaschke and Baumler 1989, Johnson 1996) and invertebrates (Fogel 1975) that disperse them, and range from nitrogen-fixing bacteria and their associates, invertebrates to detritivores, defoliators, predators etc., cavity-nesting birds and raptors. Effective conservation of biodiversity in European forests helps maintain a full set of "ecological processes" (e.g. decomposition, nutrient and hydrological cycling, succession), and hence provides for an ecosystem with high stability (Schulze and Mooney 1994). In order to maintain biodiversity in a broad sense, it is necessary to preserve or stimulate the "patterns and processes" giving rise to heterogeneity at a range of spatial scales (Halpern and Spies 1995).

Measurable processes generating and maintaining biodiversity include (after Boyle et al. 1998):

1) NATURAL DISTURBANCE REGIMES: Disturbance is a key process in natural forests of Europe. Disturbance agents may be endogenous and exogenous, from biotic or abiotic causes, and encompass a very broad range of temporal and spatial scales (Quine et al. 1999). Disturbance provides the driving force for forest dynamics and regeneration through structural change, the initiation of secondary succession and creation of habitat diversity (Pickett and White 1985, Quine et al. 1999). Disturbances of intermediate severity are a major diversifying force in forest ecosystems (Petraitis et al. 1989), and random periodic disturbances are known to maintain high species richness and productivity and limit competitive exclusion (Huston 1979). The links between site type and disturbance regime are critical because they shape the composition and structure of the forests, as well as

many of the important processes, to which forest species have adapted (Angelstam 1998a, b). The maintenance and restoration of the full range of available sites and disturbance regimes, tree and other keystone species and seral stage representation found in the naturally dynamic forest is therefore required. It also requires a diversity of stand sizes, juxtapositions and configurations as well as the pertinent processes affecting forest ecosystem maintenance (Hunter 1990, Angelstam 1992, 1997).

2) DISPERSAL/MIGRATION: The dispersal and migration of species ultimately determine future forest composition and pattern. Human interventions may affect the capacity of the landscape to provide suitable sites for dispersal or migration and hence alter the pattern of gene flow. This has been noted in studies concerned with the effects of structural change and fragmentation on the migration and dispersal of sika deer *Cervus nippon* (Staines et al. 1998); and the effects of soil compaction on dispersal (mycophagy) of below ground fruiting mycorrhizal fungi (Johnson 1996). The effect of fragmentation varies from species group to species group. For a certain number of invertebrates, e.g. some beetles even a forest path can be a obstacle for migration, while most invertebrates and also in general the vertebrates are effective migrators.

3) REPRODUCTION: Reproduction also determines future forest ecosystem composition. Impacts on the process of reproduction (e.g. decline in fecundity and an increase in mate search time as a result of landscape fragmentation) can have rapid, direct and dramatic consequences on biodiversity. In the case of species with short generation times, non-overlapping generations, or highly specific mutualisms, changes can be particularly devastating (e.g. habitat loss or simplification/even-aged forests at a susceptible stage to insect attacks).

4) REGENERATION/SUCCESSION: Gap-phase dynamics promote successional diversity and determine the natural patterns of forest community dynamics. Adequate regeneration by seed is a fundamental aspect of sustainability (Hartshorn 1995). After disturbance (e.g. felling) seral communities develop in different stages e.g. a reduction in area of mature, or "old-growth" forest, with replacement by communities dominated by pioneer, or early successional species. Forestry operations that alter ecosystem parameters beyond a critical limit are thought to cause a permanent change to the ecosystem, leading to an arrested climax (Boyle et al. 1998). It follows that the preservation of biodiversity in temperate forests requires all successional stages to be represented (Franklin 1988).

Table 1. Definition and description of the key components of biodiversity (Franklin et al. 1981).

Components of biodiversity and ecological integrity	Definition and description with reference to forest ecosystems
Composition	Composition encompasses the identity and variety of elements in a collection, and includes species lists and measures of **species** diversity and **genetic diversity**. Ecosystems are composed of organisms, species, groups of closely interacting species, genetic diversity within species, legacies of organisms (e.g. dead wood and soil organic matter), and various inorganic components (e.g. minerals and gases).
Structure	Structure is the physical organisation or **pattern** of a system, from habitat complexity as measured within communities, to the pattern or patches and other elements at a landscape scale. Ecosystem structure arises from the patterns in which components of composition occur. It is useful to distinguish two aspects of system structure: a) architectural, and b) social. Architecture denotes the physical aspects of structure, and is most often applied to spatial patterns in the plant community: the number of canopy layers, for example, or patchiness in the distribution of species and age classes. Social structure refers to patterns in the way that individuals, species, or groups of species relate to one another and to the system as a whole e.g. predation, symbiotic relationships or mutualisms.
Function	Function involves ecological and evolutionary **processes**, including gene flow, disturbances, and nutrient cycling. There are two aspects of function: a) the influence on processes (e.g. photosynthesis, nutrient cycling, population growth), and b) the influence on system structure (e.g. balance among different populations). In addition to these internal functions, we can also identify external functions, which are influences of the community as a whole on its surroundings. Regulation of water and nutrient fluxes, stabilisation of soils, and absorption and reflection of solar energy (i.e. albedo) are examples of external functions.

5) TROPHIC DYNAMICS: Ways in which species from different trophic levels interact include predation and herbivory. As each trophic level is dependent on other levels, impacts on trophic dynamics can be very serious to forest ecosystem functioning. For example, forest operations in primeval forests of eastern Europe alter forest structure and hence the carrying capacity of red deer *Cervus elaphus*, with consequential effects on the survival of predatory wolves *Canis lupus* (Jedrzejewska et al. 1994).

6) ECOSYSTEM PROCESSES: A full and operational set of ecosystem processes is essential for ecosystem functioning, and stability (the ability of the forest system either to resist change or to rebound after change; Perry 1994). Such processes include photosynthesis, nutrient and hydrological cycles, dynamic aspects of food webs, succession, evolution, migration, and the movement of disturbances across landscapes. Disturbance to, or removal of processes influence the ability of an ecosystem to function efficiently. In boreal forests for example, damage to soil fauna is thought to disrupt decomposition, mineralisation and primary production in coniferous forest soils (Huhter et al. 1998).

7) LOCAL EXTINCTION: Local extinction is the disappearance of a local population or metapopulation. It is a process rather than an event, and today is initiated by rapid human-induced environmental change (Angelstam 1998a, b). It ultimately results in the elimination of individuals that potentially contribute to ecosystem functioning. Populations which drop below their critical sizes become vulnerable to extinction because of both demographic and genetic factors, the later including inbreeding depression and loss of genetic diversity. The size of populations necessary to maintain viability varies widely depending on species and environment. Preserving adequate numbers of individuals for given species requires that sufficient suitable habitat to maintained, including all components of the ecosystem on which a given species either directly or indirectly depends (Perry 1994). Loss of keystone species (species that play a pivotal role in ecosystem (or landscape) processes and "upon which a large part of the community depends"; Noss 1991), produces cascade effects (i.e. leads to the loss of other species or the disruption of processes).

1.2 Policy development as regards the management of forest biodiversity

The interest in maintaining the forest resources has occupied European societies since pre-historic times, admittedly with little success in e.g. the Mediterranean region. To generalise, forests have to a large extent survived either in areas of little economic interest (for various reasons, including population decline) or in areas where king and nobility controlled the land-use and preserved forests for e.g. hunting or shipbuilding purposes. A great exception, where intense land-use favoured trees is agroforestry systems; olive and oak in southern Europe and, for several species in certain areas, leaves were a resource for the winter survival of the domesticated animals.

The increasing industrial exploitation of the forest in the 1800s for timber (and later pulp) created in many European countries a general concern about sustainability of the wood resource, resulting in a number of National Forest Laws during the 1900s. These Forest Laws in several countries developed into a framework of more general nature resource management, for sure also including biodiversity during e.g. the 1970s (but the word was not introduced).

On the political arena a global concern for the loss of biodiversity was introduced in the early 1990s. This was a main theme at the United Nations Conference on Environment and Development (UNCED 1992) in Rio de Janeiro. "Forestry" and "biodiversity" issues were considered a priority at the conference. As a result, a large number of European countries a) signed a framework agreement, the "Convention on Biological Diversity" (CBD), b) agreed on a set of "Forest Principles" and finally, c) contributed to a component of "Agenda 21", outlining a non-legally binding authoritative statement of principles for a global consensus on the management, conservation and sustainable development of all types of forest.

Since the CBD, various national and international initiatives, dealing with criteria and indicators including the maintenance of biological diversity in forests have been implemented world-wide (Van Bueren and Blom 1997). Among the most notable of these were the "Intergovernmental Seminar on Criteria and Indicators for Sustainable Forest Management (the Helsinki Process) (1994), the meetings of the "Intergovernmental Working Group on Global Forests" (the Montreal Process) (1995), and the "Tarapoto Proposal of Criteria and Indicators for Sustainability of the Amazon Forest" (1995).

In Europe, activities to implement the CBD have included e.g. the following pan-European/European Union activities:

• The Ministerial Conference "Environment for Europe" which started in Dobris (1990), followed by Lucerne (1993) and Sofia (1995), where the Environment Ministers of 55 European Countries and the EU Commission endorsed the "Pan-European Biological and Landscape Diversity Strategy" (PEBLDS) (1996–2000).

• The Ministerial Process "Ministerial Conference on the Protection of Forests in Europe (MCPFE)" which started in Strasbourg (1990) and was followed by the Helsinki Conference (1993) and Lisbon Conference (1998). Its resolutions were signed by 37 European countries and the EU Commission. Especially the Resolution H2 "General Guidelines for Conservation of the

Biodiversity of European Forests" focus on the regional implementation of the CBD.

- The "Community Biodiversity Strategy" developed by the EU Commission.
- The implementation of EU "Habitats Directive" 92/43/EEC – "NATURA 2000".
- The "Community Forestry Strategy for the European Union", developed by the EU Commission.
- The European Environment Agency EEA set up as an independent organisation, but with all EU countries as active members, with the role of organising environmental monitoring in the EU. A strong supportive structure has been created in the form of Topic Centers; for biodiversity most notably the Topic Center Nature coordinated by the Museum of Natural History in Paris. EEA (with the aid of its Topic Centers) has already produced two extensive reports on the State of the art of European environment.

Similarity in the objectives expressed in the PEBLDS (Action theme 9 – Forests) and the Helsinki resolution (H2) lead to the preparation of the "Work Programme on the Conservation and Enhancement of Biological and Landscape Diversity in Forest Ecosystems" (WP-CE-BLDF) (1997– 2000). This was endorsed in June 1998 by Ministers of the Environment in Århus, and by Ministers of Forests in their Third Ministerial Conference, in Lisbon, cf. Rego and Dias (1999).

Throughout Europe, regional and national initiatives have also been drafted (or are being researched), and in most cases implemented. These contribute towards fulfilling the national obligations set out in the CBD and include e.g.:

BELGIUM (FLANDERS REGION): Forestry long-term planning (Ministry of the Flemish Community) implemented by the "Action plan for Forestry" in the framework of the "Environmental Policy and Nature Development Plan (MINA plan 1997–2001)".

DENMARK: Strategy for Natural Forest and other Forest Types of High Conservation Value in Denmark (Ministry of the Environment, 1994), Strategi for bæredygtig skovdrift (Ministry of the Environment, 1994), Biologisk mangfoldighed i Danmark – Status og strategi (Ministry of the Environment 1995).

FINLAND: Maa-ja metsätalousministeriö. Suomen kestävän metsätalouden kriteerit ja indikaattorit metsätalouden tilan kuvaajina (1997).

FRANCE: Les indicateurs de gestion durable des forêts françaises. Ministère de l'agriculture et de la pêche, France.

GERMANY: Federal and Regional laws. BMELF (Federal Ministry of Food, Agriculture and Forestry/Bundesministerium für Ernahrung, Landwirtschaft und Forsten) (1999): Forstwirtschaft und Biologische Vielfalt. Strategie zur Erhaltung und nachhaltigen Nutzung der Biologischen Vielfalt in den Offentlichen Waldern Deutsch-

lands. Sturm, K. (1993). Prozasschultz – ein Konzept für naturschutzgerechte Waldwirtschaft. Z. Ökologie u. Naturschutz 2: 182–192.

GREECE: New forest policy in management plans for conserving and increasing biodiversity and particulary in coppicing-managed forests (oak, chestnut) (Ministry of Agriculture – General Directorate of Forests and Natural Environment, Greece, 1998).

IRELAND: National Standard of sustainable forest development in Ireland. Draft, February 2000. Edited by J. J. Gardiner and B. McGuire. Coillte Forests: A vital resource: A framework for sustainable forest management. May, 1999. Irish Forestry Board, Dublin.

ITALY: Intermediate research report: Individuazione di un sistema d'indicatori di gestione forestale sostenibile in Italia. Rapporto intermedio di ricerca. 54/1999. Gennaio 1999. Agenzia Nationale per la Protezione dell'Ambiente e Dipartimento Territorio e Sistemi Agroforestali, Università di Padova (Bortoluzzi, B., Fedrigoli, M., Pettenella, D. and Urbinati, C. (1999)). Documento informativo di Supporto alle Organizzazioni Forestali per l'Utilizzo delle Norme UNI EN ISO 14001 e 14004 sui Sistemi di Gestione Ambientale. Septembre 1999. Ente Nazionale Italiano Unificazione (UNI), Gruppo di Lavoro 9 "Gestione sostenibile delle foreste" and Pettenella, D. and Secco, L. (Dipartimento del Territorio e Sistemi Agroforestali Universit di Padova).

MADEIRA: Action Plan for Madeiran Sustainable Forest Management (in preparation since 1998).

PORTUGAL: DGF (1998): Plano de Desenvolvimento Sustent vel da Floresta Portuguesa. DirecUo-Geral das Florestas. Ministrio da Agricultura, Desenvolvimento rural e pescas. Research programmes (e.g. PAMAF), The National Law for Forestry Policy, and National Forestry Inventories have also incorporated forest biodiversity issues.

SLOVENIA: The Forest Development Programme of Slovenia. Ministry of Agriculture, Forestry and Food. Republic of Slovenia (1996).

SWEDEN: Action Plan for Biological Diversity and Sustainable Forestry (The National Board of Forestry, Sweden, 1996).

U.K.: The U.K. Forestry Standard: The governments approach to sustainable forestry (The Forest Authority, Forestry Commission, U.K., 1998).

1.3 Biodiversity assessment and evaluation

Throughout international policy, the identification of indicators for the assessment and evaluation of biodiversity has been stressed. In the European context, the first action proposed by the Work Programme on the Conservation and Enhancement of Biological and Landscape Diversity in Forest Ecosystems (WP-CEBLDF) was indeed the identification of indicators for assessing biodiversity in European forests. Biodiversity assessment and evaluation is necessary to ensure management prescriptions or that strategies are in harmony with international and national expectations, and to ensure that forest management practice standards (e.g. certification standards) are being met.

Concern for the future of forest biodiversity has generated environmentally friendly forestry initiatives such as certification, or the "green labelling" of forest products. Certification is the prime means of providing the economic incentive needed to promote the sustainable management of and biodiversity conservation in forests (Bass 1998). Certification programmes call for voluntary compliance with established environmental standards in exchange for higher prices or greater market access or both.

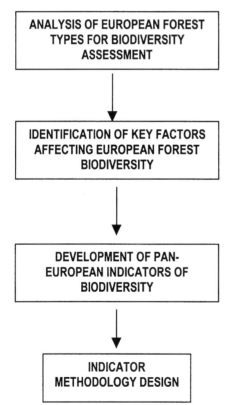

Fig. 1. Summary of the hierarchical logic involved in the development of BETs (indicators of biodiversity and the methodology needed to apply them).

The certification scheme developed by the Forest Stewardship Council (FSC) is by far the most visible on the international scene. National programmes such as the U.K. Woodland Assurance Scheme (UKWAS), the Finnish Forest Certification System (FFCS) etc. operate in association with FSC. The launch of the "Pan-European Forest Certification Program" (PEFCC) in June 1999, highlights the commitment towards certification (PEFCC 1999).

Implicit to certification and other environmentally or conservation friendly practices, is the assumption that we can assess and evaluate the improved sustainability/conservation value of the forestry practices in question (Kimmins 1997b).

It is not possible, however, to easily measure and quantify the whole spectrum of biodiversity in a comprehensive manner. For an evaluation measurable parameters are needed which act as correlates or surrogates for intuitively "true biodiversity" (Duelli 1997). It is therefore highly desirable to have indicators or surrogates, at least for certain aspects or dimensions of biodiversity (e.g. composition, structure and function). These indicators require 1) adequate measurement and 2) pertinent interpretation to ensure a successful evaluation of biodiversity (Duelli 1997). It is also important that we have a thorough understanding of natural processes and biodiversity characteristics/components in undisturbed forests (Standovár 1996) to act as natural standards for comparisons.

Biodiversity Evaluation Tools BETs are the guidelines for biodiversity assessment; i.e. a suggested indicator scheme and optional methodologies for data collection. Figure 1 presents the principles used to develop BETs for European forests.

1.4 The EU project "Indicators for monitoring and evaluation of forest biodiversity in Europe BEAR"

The project "Indicators for monitoring and evaluation of forest biodiversity in Europe BEAR" was initiated in 1998, under the auspices of the European Commission Agriculture and Fisheries RTD programme (Contract FAIR5 CT97-3575). The BEAR project is a pan-European concerted action, bringing together expertise from 27 research organisations of 18 European states to build a framework for the development of forest biodiversity indicators at various scales, cf. Table 2, Appendix 1.

The BEAR project thus aims to present a framework for the development of Biodiversity Evaluation Tools (BETs) which can be defined as the combination of biodiversity indicators and the standardised methodology needed to apply them. The work procedure to present the framework for BETs has included:

• An analysis of European forests with respect to key fac-

Table 2. The BEAR project organisation and participants (more information in Appendix 1).

Function in BEAR	Organisation	Responsible scientists
Co-ordination	Swedish Environmental Protection Agency	Tor-Björn Larsson, *co-ordinator* Kjell Sjöberg, *assistant co-ordinator* Linus Svensson & Gunilla Andersson, *reporting and webservice*
Mediterranean and Macaronesian regional group	Estação Florestal Nacional, Portugal	Francisco Rego, *regional co-ordinator* Susana Dias, *assistant regional co-ordinator*
	Jardin Botanico da Madeira	Fransico M. Fernandes
	Facultad de Ciencias del Medio Ambiente Universidad de Castilla-La Mancha, Spain	Federico Fernández-Gonzalez
	Academia Italiana di Scienze Forestali	Orazio Ciancio, *scientific responsible* Anna Barbati Piermaria Corona Marco Marchetti, *contact* Roberto Scotti
	NARF Forest Research Institute, Greece	Konstantinos Spanos
Atlantic regional group	Centre for Ecology and Hydrology, Banchory, U.K.	Allan Watt, *regional co-ordinator*
	Forestry Commission Research Agency, U.K.	Richard Ferris Susannah Hall
	University College Cork, Ireland	John O'Halloran Paul Walsh Paul Giller
	Wageningen Univ. and Research Center & Alterra, Green World Research, The Netherlands	Frits Mohren Rienk-Jan Bijlsma
	Instituut Voor Bosbouw En Wildbeheer, Belgium	Kris Vandekerkhove Diego VanDenMeersschaut
	Laboratoire d'Entomologie Forestière, INRA, France	Hervé Jactel
	INRA–SAD Toulouse, Terrestrial Ecology, France	Gérard Balent
	Geological Survey of Denmark and Greenland	Richard Bradshaw Peter Friis Møller
Continental and alpine regional group	Forstliche Bundesversuchsanstalt, Austria	Georg Frank, *regional co-ordinator*
	Forstliche Versuchs- und Forschungsanstalt Baden-Württemberg, Germany	Winfried Bücking
	Federal Research Centre for Forestry and Forest Products, Germany	Herrmann Ellenberg
	Universität Göttingen, Germany	Klaus Halbritter
	Universität für Bodenkultur, Austria	Ewald Rametsteiner
	Swiss Federal Institute for Forest, Snow and Landscape Research	Christoph Scheidegger John Innes Michéle Keannel Dobbertin
	Eötvös Loránd University, Hungary	Tibor Standovár
	University of Ljubljana, Slovenia	Jurij Diaci Andrej Boncina
Boreal regional group	Norwegian Institute for Nature Research	Bjørn Åge Tømmerås, *regional co-ordinator*
	European Forest Institute	Janne Uuttera
	Unversity of Joensuu, Finland	Pekka Niemelä
	Swedish University of Agricultural Sciences	Per Angelstam Lennart Hansson
	The Forestry Research Institute of Sweden	Lena Gustafsson

tors of biodiversity at national, landscape and stand level and a subsequent identification of 33 European-level Forests Types for Biodiversity Assessment FTBAs;
• Presenting a strategy for the development of indicators of forest biodiversity to assess the key factors most relevant to each FTBA including methodological considerations;
• In dialogue with most important user-categories give recommendations with respect to development of BETs adapted to each specific use.

The BEAR project received 216 000 euro from the EU Commission which allowed the experts during the two-year project period to hold three all-group meetings and at least one regional meeting to discuss the work. Most of the work was performed through internet communication. A specific user-panel was established, cf. Appendix 1, with which a systematic consultation of tentative ideas and results was performed at two occasions.

Acknowledgements – The BEAR project has received a great interest from the scientific community, policy-makers, forest managers and others. The project participants acknowledge most stimulating discussions in connection with individual correspondence, presentations and seminars. The contacts with the user-panel, cf. Appendix 1, have resulted in a most valuable guidance for the project work. Additionally the following specific contributions should be acknowledged: Udo Bohn, Bundesamt für Naturschutz BfN provided the map of Potential Natural Vegetation which was the basis for the identification of the Forest Types for Biodiversity Assessment. We also acknowledge the general interest of BfN in the BEAR project, acting as a "28th partner" through the active participation of Manfred Klein. A co-ordination team was created through support funding by Swedish Environmental Protection Agency and the Swedish Council for Agricultural Research: Kjell Sjöberg, Swedish University of Agricultural Sciences assisted in co-ordinating the project, Linus Svensson/ZooBoTech HB managed the website and produced the Newsletters and Technical Reports. Furthermore Linus Svensson collected the descriptions of the FTBAs and together with Gunilla Andersson edited this report, including drafting of maps. Susannah Hall Forestry Commission Research Agency, U.K. was a great help in a critical stage of reporting and producing the text for the introductory chapter.

Chapter 2.

Biodiversity in European forests

2.1 Historical development of the European forests

The unique characteristics of European physical geography and cultural history have influenced the long-term development of forest vegetation and biodiversity to make them distinct from the richer faunas and floras of other continents in the Northern Hemisphere. At the end of the Tertiary, ca two million years ago, the woody floras of the Northern Hemisphere contained less inter-continental variation than today, as far as can be judged from the fossil record. The last two million years were characterised by the rapid climate changes of the Quaternary glaciations that exerted enormous pressure on European forests. For most of this time, arctic or subarctic climates caused non-forest conditions to prevail in northern Europe. The climate of north-west Europe is closely linked to the temperature of surface waters in the north-east Atlantic Ocean and these can change rapidly. Sudden falls in temperature in the past led to the growth of continental ice sheets and swift, massive regional extinctions of entire forest communities, which only survived in locally favourable refuge areas. The present forest biodiversity of Europe is largely determined by the nature of these glacial refugia. These were primarily located at the southern and south-eastern margins of Europe in areas of varied relief and soil type. Forest vegetation was probably confined to fragmented, narrow belts in the mountains during the periods of harshest climate. Survival downslope was restricted by drought conditions and upslope by alpine conditions. The refugia were small, scattered and isolated from each other and many species disappeared from Europe by the same process of habitat fragmentation and local extinction that occurs today as a result of anthropogenic impact. The floristic impoverishment that occurred in these refugia resulted in less and less species re-colonising north-west and central Europe during the brief inter-glacial episodes such as the present one. This loss of biodiversity was much subdued in other continents where the refugia were less isolated and larger, and where the climatic gradients were less severe.

The last ten thousand years of the present inter-glacial saw a new development in this pattern of gradual loss of European forest biodiversity. Humans began to influence their environment on a regional scale, firstly through hunting large mammals to the point of extinction and subsequently through the spread of farming. We can follow forest development in some detail thanks to the network of fossil pollen sites that have been studied and mapped (Fig. 2). These show a mosaic of mixed forest 9000 yr BP developing into a predominance of deciduous forest 6000 yr BP, at the time of higher summer temperatures and a longer growing season. Boreal forest with *Picea abies* was developed in the alpine region, but *Picea* was largely absent from Scandinavia. Deciduous forest was widespread in the Mediterranean, with the typical fire-prone woody scrub confined to the drier eastern regions. The last 6000 yr are characterised by the spread of *Picea* and *Fagus sylvatica* and a retreat of the rich, deciduous forest characterised by *Quercus, Tilia* and *Corylus*. Most of these changes are thought to reflect continuous climatic change, but human influence, particularly the spread of agriculture from south-east to north-west was of continuously increasing importance. The mass expansion of *Fagus* in particular has been closely linked to human disturbance, as has a part of the decline in importance of deciduous forest. The westward spread of Mediterranean scrub is apparently a consequence of both increasing aridity and anthropogenic impact including husbandry of goats and sheep, tree-cutting and vegetation management using fire.

Forest biodiversity is linked to the dynamics of the dominant trees, but forest structure and disturbance regime are other contributory factors. Fossil pollen studies suggest that natural forests had a generally closed structure and were not similar to modern wood-pasture earlier in the Holocene, despite the presence of some large herbivores. Frequent and widespread forest fires and the fall of giant trees created some sun-exposed habitats and an abundance of dead wood. The more recent spread of the shade-tolerant and shade-giving *Picea* and *Fagus* together with the large-scale withdrawal of domestic animals from forest

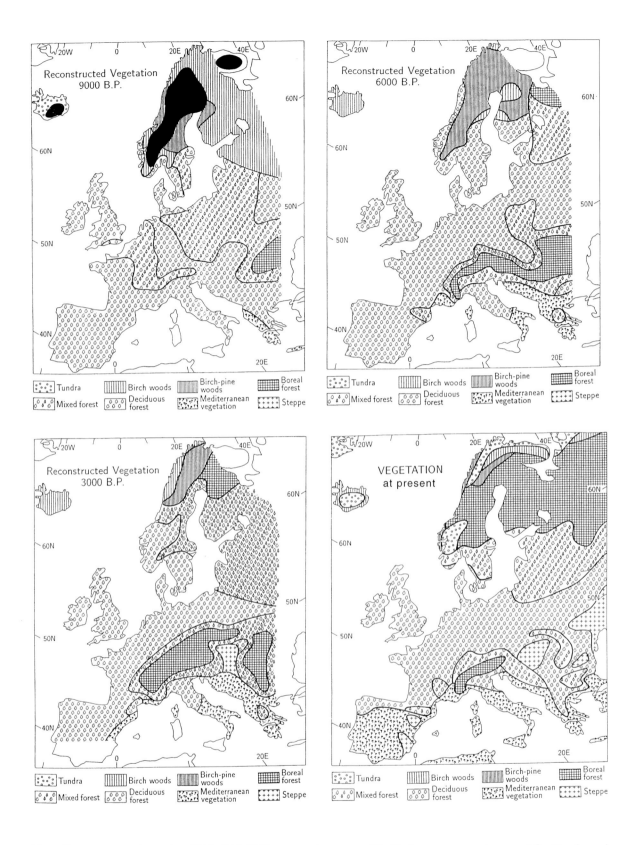

Fig. 2. European vegetation regions 9000, 6000, 3000 yr ago and at present. The black area represents ice cover (after Huntley and Prentice 1993).

habitats and widespread adoption of intensive silvicultural management have made current forests denser and darker than before. Throughout Europe the present-day forests are heavily influenced by human activities now and in the past. Plantations of introduced species dominate forests in some parts of Europe. Very few or even no forest areas can be regarded as strict virgin forest. The transformation of natural forests and the extinction of virgin forest began in south-eastern Europe with agriculture development and exploitation over 10 000 yr ago and was more or less complete 5000 yr later. The economic exploitation of European forests developed from simple timber extraction to silvicultural management at different times. Modern silvicultural principles were introduced into central Europe 400–500 yr ago and reached northern Scandinavia during the 1900s. Afforestation programmes are known to have occurred locally since at least the 1600s, but were only nationally organised during the 1900s. Awareness of biodiversity aspects in forest management can already be seen in the works of John Evelyn during the 1600s, but only gradually gained a more general acceptance during the last 150 yr.

The present fragmented forest condition in Europe has some similarities to the situation during glaciations. However the critical refuge areas are also under anthropogenic pressure, and loss of species here is potentially disastrous. The present combined pressures of rapid climate change and extreme anthropogenic exploitation have placed European forest biodiversity in a critical condition.

2.2. The state of biodiversity in European forests

The European continent has nearly 215 million ha of forests (and other wooded land) which is nearly 30% of the continent land area (EU Commission 1998). Although Europe contains only a fraction (ca 6%) of the total forest land the range of forest types is relatively vast. Forests vary from Mediterranean evergreen broad-leaved forests to temperate deciduous forests in central Europe, through to boreal coniferous forests in the north (Mayer 1984, FAO 1999).

The variation in potential natural vegetation (and the diversity of life supported by it) is a result of the heterogeneous nature of geology, topography and climate, and the consequence of post-glacial floral and faunal recolonisation. As a result, habitat classification is complex (Noirfalise 1987), and is further complicated by more or less intense agricultural practices and other anthropogenic disturbance, which have had a major impact on the development of European nature since the Bronze age.

The main overall EU initiative to preserve biodiversity, NATURA 2000, e.g. playing a key role in connection with the implementation of the habitats directive (Council Directive 92/43/EEC), builds on several attempts to classify European biodiversity. For the development of Biodiversity Evaluation Tools for European forests the following European classifications should be taken into account:

Table 3. Forest cover in the 18 European countries participating in BEAR (after Parviainen et al. 2000).

Country	Area of forest (1000 ha)	Forest as percentage (%) of total land area
Austria	3 924	47
Belgium (Fland.)	135	10
Belgium (Wall.)	530	31
Denmark	445	11
Finland	23 000	76
France	15 156	28
Germany	10 700	30
Greece	6 513	49
Hungary	1 748	19
Ireland	600	9
Italy	8 675	29
The Netherlands	334	10
Norway	11 950	37
Portugal	3 306	37
Slovenia	1 110	54
Spain	12 511	25
Sweden	28 000	69
Switzerland	1 186	29
U.K.	2 305	10

1. EUROPEAN BIOGEOGRAPHIC REGIONS: A relatively simple division of Europe into six major biogeographic regions was introduced in connection with the habitats directive, see Fig. 3. The biogeographic regions reflect major differences in biodiversity caused by climatic and other site differences (in combination with historical factors). This classification is introduced here to promote thinking in more nature-based units for biodiversity management than the countries (that are hitherto used as operational units), cf. Chapter 6.1. The 6 regions are not further described in e.g. the NATURA 2000 interpretation manual (European Commission 1996). They can be characterised and described roughly:

The BOREAL REGION covers the northernmost parts of Europe and is characterised by both a north-south and a west-east climatic gradient. The winters are in general long and cold with coherent snow coverage. The W–E gradient is from very oceanic conditions with high precipitation (>2000 mm) along the west coast of Norway to dry and continental conditions (<400 mm) in Lapland, particularly in northern Finland. The summer temperatures along this gradient are highest in the east (16–18° in July) and lowest in the west (12–14° in July). Increasing geographic latitude implies shorter growing season northwards.

Except for the mountainous regions where birch *Betula* spp. is dominant just below the tree-line, *Picea* and *Pinus* are dominant species. This prevailing coniferous forest often occurs in a complex mosaic with bogs and mires. The soils are generally young due to coverage of ice during the Weichselian glaciation, but acid due to the bedrock. The amount of species in the forest systems is increasing southwards.

The ATLANTIC REGION covers Ireland, U.K. and the west coast of the European continent. The region is influenced by the vicinity to the Atlantic Sea, causing a humid oceanic climate with a low degree of annual variation and in general a high precipitation. The winters are in general mild and the summers cool. The forest percentage in the region is the lowest in Europe, due to exploitation. The natural forests of the region are mainly broad-leaved, but to a large degree replaced with conifer plantations. The species diversity is in general low.

The CONTINENTAL REGION covers most of central and eastern Europe. The climate is less influenced by the sea and the winters are colder and summers drier and warmer, causing an E-W gradient and distribution pattern for many species. Only minor parts of the region were affected by the Weichselian glaciation and the soils are in general old.

The ALPINE REGION covers the most mountainous parts of Europe and is of course characterised by the alpine conditions causing a significant zonation with a tree-line from 1000 – 2000 m a.s.l. The timber- and tree line is formed by conifers (pine species) in central Europe. The alpine region have some similarities with the boreal region but the solar influx is much higher and the species diversity several times higher.

THE MEDITERRANEAN REGION covers most of the Iberian Peninsula, the Mediterranean islands, Italy, Greece and the coastal zone of the Balkan Peninsula. The climate is warm and in general dry, and in large parts precipitation is a limiting factor. Frost only occurs occasionally in the lowland in wintertime.

The forests have been intensively influenced by man for thousand of years with cuttings (to some degree coppice), fires, grazing (particularly goat grazing) and deforestation. In large parts remnants of the natural forest consist of species rich shrub with e.g. several oak species including the sclerophyllic *Quercus ilex* and *Q. coccifera*.

THE MACARONESIAN REGION consists of the small, mainly volcanic islands in the Atlantic Ocean west of Africa with – except for the high mountainous parts – a more or less subtropic climate where frost is extremely rare. As a result of a long period of isolation due to the continental drift, the flora contains several relics from the Tertiary period and a huge amount of endemic species. The richest forests are found in gullies and north oriented slopes.

2. IMPORTANT EUROPEAN FOREST TYPES ACCORDING TO THE HABITATS DIRECTIVE: The European Habitats Directive (Council Directive 92/43/EEC) lists a large number of important forest types for which European Governments are required to "maintain their favourable conservation status", see Appendix 2. These forests are listed in Annex 1 of the Directive that specifies that their conservation requires the designation of SAC (Special Areas of Conservation). The forest types (of tall trees) listed are "rare or residual, and/or hosting species of community interest". A number of forest types are treated as priority habitat types (e.g. Caledonian forest, Macaronesian laurel forests, Mediterranean pine forests with endemic black pines, *Taxus baccata* woods). Additional types of "low forest or scrub" are also listed under other habitat types in Annex 1 (with priority types including wooded dunes with *Pinus pinea* and/or *Pinus pinaster*). Other forest types, such as the plantation forests of Britain and Ireland provide important habitats for species of high conservation importance including the nightjar *Caprimulgus europaeus* (Annex 1 Birds Directive).

3. CORINE BIOTOPES: The CORINE Programme was in early 1990s set up by the Commission to study possibilities for CO-oRdination of INformation on the Environment. The CORINE Programme thus formed a background for the European Environment Agency to start its work. At the same time the site description concept as well as the CORINE Habitat Classification formed the initial basis for the Commission work on NATURA 2000. The CORINE Biotopes information has been used very differently by Member States as a source of information for the national NATURA 2000 Network processes.

4. EEA/CORINE LAND COVER: The European Environmental Agency (EEA) was established in 1990 by the Eu-

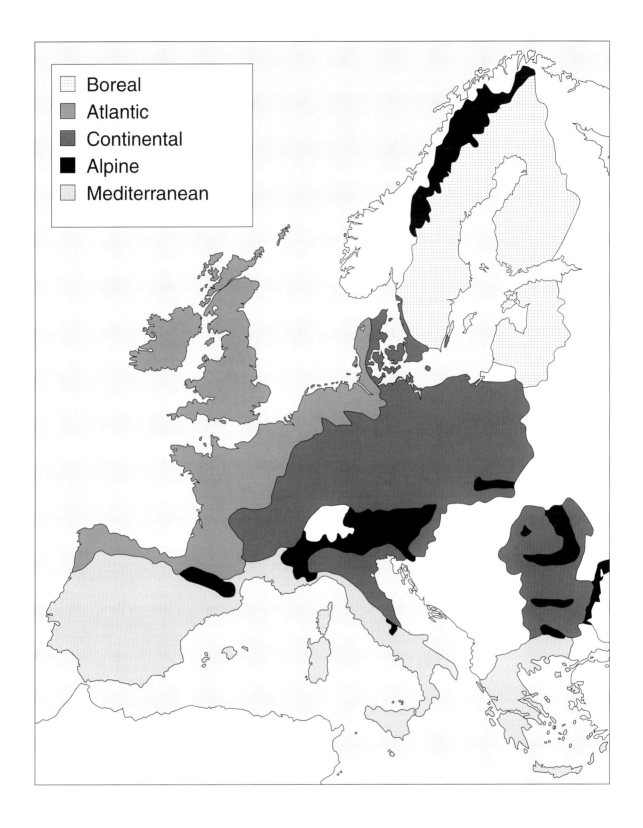

Fig. 3. The European Biogeographic Regions as identified in the Habitats Directive, cf. text. Note: The Macaronesian Region (outside map): Madeira, Azores, Canary Islands.

ropean Council to provide the Community, to member States and to other European countries with comparable information at European level, enabling them "to take requisite measures to protect the environment, to assess the results of such measures and to ensure that the public is properly informed about the state of the environment". Regarding land cover, EEA decided to compile a land cover data base (CORINE) of all European countries by using computer assisted image interpretation of all observation satellite images. A standard methodology and nomenclature were agreed on resulting in three different levels, allowing to distinguish 44 different types of land cover, that can be further supplemented with more detailed levels if individual countries so desire. A description of CORINE Land Cover types as regards forests is given in Appendix 3.

5. POTENTIAL NATURAL VEGETATION: Still another approach to describing European biodiversity has been used as a basis for development of Biodiversity Evaluation Tools for European forests: The map and units of Potential Natural Vegetation developed in the project "European Vegetation Survey" (Bohn 1994, 1995). This project has included a cooperation of some 30 Institutes in 30 European countries since, at least the early 1990s. A major European level output has been a map at a scale of 1:10 million including a hierarchical legend of types of potential natural vegetation, i.e. "main natural plant communities corresponding to actual climatic and edaphic conditions, excluding – as far as possible – human impact" (Bohn 1995 p. 144). In the BEAR project we found the map of Potential Natural Vegetation to be the best available European level basis for delimiting the Forest Types for Biodiversity Assessment FTBAs, cf. Chapter 3.2 and the maps presented for each FTBA in Chapter 4. The legend of the European map of Potential Natural Vegetation is presented in Appendix 4.

European forest biodiversity is of significance e.g.:

- As part of cultural landscapes (e.g. medieval pasture woodlands in the U.K. and Ireland, "dehesas" of Spain, "coltura mista" in Italy and areas with old age oak trees in different parts of Europe – e.g. Greece, Sweden);
- Aesthetically for their beauty (e.g. birds, butterflies and flowering plants);
- For their "direct economic worth" e.g. food, timber, non-timber forest products or land value for conversion (Ehrlich and Ehrlich 1992);
- As a vital habitat, which supports diverse plant and animal communities, providing refugia for rare, endangered and/or sensitive species (e.g. brown bear *Ursus arctos* in Greece; dormouse *Muscardinus avellanarius* and red squirrel *Sciurus vulgaris* in the U.K.);
- Providing vital ecosystem services e.g. as regards carbon sinking (reduction of global carbon dioxide levels), water and nutrient balance, soil protection etc.

It is mainly their direct economic worth as regards wood resources that during last decades has threatened the biodiversity of European forests. In a more long-term perspective it was the transformation of forest land due to agriculture and grazing, including burning, draining and extermination of wild competitors (large predators, beaver *Castor fiber*, wild boar *Sus scrofa* etc). Introduction of alien species has in European forests hitherto created significant (e.g. *Eucalyptus, Rhododendron, Robinia*) albeit relatively moderate problems from a biodiversity point of view. Forestry measures, mainly as regards selection of forest tree seed, have also in Europe to a relatively large, but not fully documented, extent greatly impacted the genetic composition of the tree populations.

Dramatic changes in forest ecological conditions and habitat development have occurred throughout northwest (Bradshaw et al. 1997, Björse and Bradshaw 1998, Bradshaw and Holmqvist 1999, Bradshaw and Mitchell 1999) and central (Ellenberg 1986) Europe during the last 2 000 yr (see Chapter 2.1). These have been driven partly by climatic change, but of greater significance has been the transformation of forest area for agricultural or urban use, and the consequential disruption and alteration of natural disturbance regimes. Alterations of grazing regime, for example, have been particularly severe and rapid in deciduous forest areas, whilst fire management has had greater impact in coniferous regions (Bradshaw 1998) and evergreen sclerophyllus broad-leaves in Mediterranean. No truly natural forests remain in Europe apart from those in small areas which are poorly accessible, and/or where extreme climatic or topographic conditions prevail (e.g. mountainous regions in northern Europe, Bücking 1998).

Anthropogenic influences have thus significantly affected the biological diversity of European forests. Since trees as a rule are keystone species, creating specialised habitats for many dependent species, many large forest mammals and birds, particularly top predators and forest specialists have disappeared from large parts of Europe as a result of forest degradation and environmental pollution (the disappearance of the large forest mammals and top predators was, however, in its final phase caused by hunting). Many forest specialists are endangered e.g. woodpeckers, fungi and lichens (Hansson 1992, Esseen et al. 1992, Berg et al. 1994, Mikusinski 1997) but the causes are complex and not always well studied.

Nearly 3 million ha (1.6% of total forest area) of "natural" forests exist in Europe (except Russia) in strict forest reserves or other protection categories. There are over 3500 strict forest reserves in European countries (Parviainen et al. 2000). The majority of European forests are subject to anthropogenic/cultural disturbance (e.g. forest management, road construction, tourism, unnatural forest fires, livestock grazing, hunting, extraction and mining of mineral resources, drainage, water regulation and even war). These factors, often in combination, threaten the biodiversity of European forests. The ultimate result of such disturbance is the fragmentation, degradation and/or complete elimination of forests. The foremost influence is an increased demand for biological resources due to escalating population growth and economic development (Nowicki et al. 1998).

Present threats to European forest biodiversity include (after Nowicki et al. 1998):
• Fragmentation of woodlands by forest tracks, highways and electrical power wires;
• Fragmentation of agricultural land (depending when your reference point is!);
• Introduction of non-native species;
• Large plantation of conifers, with consequent acidification of soils (exception: acidification is higher in beech forest) and more frequent harvesting, excluding old trees and thus the associated flora and fauna;
• Large plantations of *Eucalyptus*;
• Increased airborne pollution;
• Increased occurrence of fires, especially in the Mediterranean region, due to increased biomass in the forests and lack of management in mainly pine and sclerophyllus forest (maintenance of undergrowth or collecting firewood);
• Increased tourism, especially in periurban forests.

Two threats should be considered more in detail 1) fragmentation; 2) forest management practices:

1) Fragmentation

Loss of natural habitats including effects of increasing fragmentation – i.e. decreasing habitat fragment size and increasing isolation between fragments (Andrén 1997, see also Kaennel-Dobbertin 1998a, b) – are major threats to biodiversity (Andrén 1997). Fragmentation effects add up to the negative effects of mere loss of habitat area. Where a keystone species has been affected by fragmentation, ripple effects may be noticed in other species not directly affected by the changing landscape (Angelstam 1997).

Forest fragmentation has a number of components (Kouki and Löfman 1998):
• HABITAT ISOLATION. The main effect of fragmentation is "geographical" – where formerly continuous areas of forest are broken into successively smaller patches. This production of habitat islands (Godron and Forman 1983) means an inevitable loss of "landscape connectivity", i.e. the prerequisites for dispersal of plants and animals between suitable habitats of the landscape;
• EDGE EFFECT. Habitat alteration takes place at forest edges, where the adjacent habitat effects the habitat quality by changed abiotic and consequently biotic factors (cf. e.g. Angelstam 1992);
• HABITAT ALTERATION, when "edge effects" cover also the forest interior habitat, resulting in more or less pronounced alteration or destruction, at least temporarily.

Critical thresholds of landscape fragmentation at which subpopulations of individual species are likely to become viable (moving from worse to better situation) or extinct (unfortunately the more common situation) will vary between species (Ferris and Purdy 1998). Fragmentation of European forests has forced larger carnivores and herbivores, which have large home ranges and seasonal movements, to move beyond forest reserve protection and into competition with farmers (Peterken 1996). Even very small-scale habitat fragmentation can have an impact on some species, cf. e.g. reports of roads creating barriers to dispersal of some animals (Mader 1984). The sensitivity to fragmentation varies from species group to species group. For beetles even a forest path can be a hinderness for migration, while e.g. birds are effective migrators (except for a few species of woodpeckers).

Fragmentation leads to a loss of core habitat area. Consequently, patch-interior species that cannot tolerate edges will become prone to extinction in small patches (Wilcove et al. 1986) (e.g. specialist carabid beetles in coniferous forests have been affected by fragmentation, as they are totally dependent upon contiguous forest cover (Halme and Niemelä 1993)). However, the effects of forest fragmentation are species specific and therefore do not always produce isolated populations (Ferris and Purdy 1998). Loss of great spotted woodpeckers *Dendrocopos major* from the Irish landscape is an important consequence of fragmentation.

2) Forest management practices

Forest management practices often have a major impact on biodiversity, causing changes to site conditions, tree species composition and forest structure (Mitchell and Kirby 1989). In Finland for example, 43% of all species classified as endangered are thought so as a result of forest practices (Kouki and Niemelä 1997), and in Denmark, bird densities in managed stands were between a quarter and a third of the density found in unmanaged stands (Møller 1997). Some of these species elements are likely to be critical to the productivity of the forest itself.

The potential effects of forest management on biodiversity occur at several scales. At the stand scale for example, removal or destruction of important habitat structures, such as coarse woody debris, during traditional boreal forest clear-cutting may affect species (Burschel 1992, Östlund 1993). At the landscape scale, fragmentation, alteration and loss of previously continuous habitat (e.g. natural old growth forests, forest fire areas, areas with a long continuity of decaying wood and old broad-leaved trees) may cause local extinctions and hamper the recolonization of maturing sites by old growth specialists (Niemelä 1999).

The major effect of forest management is a general reduction in stand age (Christensen and Emborg 1996, Esseen et al. 1997). This directly affects the diversity of organisms restricted to old trees and woody debris (e.g. insects, fungi and hole-nesting birds), whilst organisms primarily related to the "innovation" phases (e.g. vascular plants) are less affected. The light open conditions at the forest floor after thinning or clearcutting may even be beneficial for these organisms (Christensen and Emborg 1996).

Indeed, in new afforestation schemes, which by definition do not have an established woodland flora and fauna, lack of management might lead to a uniform structure in the short and medium term (Peterken 1993). In such forests, habitat diversity can be created through management and hence increase the diversity of wildlife. The provision of variety (including old growth, riparian habitats, etc.) will enhance the vital ecological processes required to provide biodiversity (Ratcliffe 1993). Managing landscapes for diversity involves managing patterns of succession. Some successional stages have more species that others; and each stage has a different, although not usually unique, set of species (Hunter 1990).

Forests can be managed to host levels of biodiversity, including important keystone species, approaching that found in natural and semi-natural woodlands in similar biogeoclimatic zones (Ratcliffe 1993). Wildlife will normally benefit from increased structural diversity for example. Indeed a varied woodland structure, together with numbers of dead trees, will allow bird life of conifer stands to approach that of broad-leaved woodland (Currie and Bamford 1982). Maintaining the continuity of ancient woodland biodiversity for example might involve management for old growth trees and high densities of deadwood, and hence lead to the protection of few endangered/specialist species such as woodpeckers, saproxylic invertebrates (Speight 1989) and fungi. Many species (e.g. butterflies) have indeed become reliant on forest management.

Forest management can therefore be considered to have both positive and negative effects on biodiversity, of course depending on the prerequisites and policy objectives for the area in concern.

Chapter 3.

Identifying main features of European forest biodiversity

3.1 Key factors of forest biodiversity

The first step in the BEAR project on the road to develop biodiversity indicators was to perform an "Analysis of important forest types in the six major European Biogeographic Regions with respect to the structure and function of major forest types in order to identify key factors of biodiversity". The idea behind this is that the present knowledge should allow us to rather well identify the key factors of biodiversity of European forests, while the assessment of these key factors, through indicators, will need further development, validation and to finally be selected by different user groups with respect to standards, available techniques and other preferences.

Within the context of the BEAR project key factors of forest biodiversity may be defined as factors that have a major influence on or directly reflects the variation in biodiversity within European forests. Key factors can be classified according to the different ecosystem components discussed in Chapter 1.1, cf. Table 1:

• Structural (physical characteristics);
• Compositional (the biological component; i.e. species);
• Functional (abiotic/biotic disturbance factors and management).

Furthermore, as discussed in Chapter 1.1, geographical scale is crucial when assessing biodiversity. To reach the BEAR objectives to deliver guidelines for assessing and manage biodiversity of European forests the following scales must be considered:

• National/regional scale (relevant for national overview and international reporting of the state of biodiversity);
• Landscape scale (see Appendix 6 for definition of landscape; as a rule a number of forest holdings comprising a variation in forest habitats and stand developmental stages; large enough to be a basis of a certain population size of more area-demanding species like many birds and mammals);
• Stand scale (forest management unit, in principle defined by the silvicultural programme).

It must be stressed that the above classification of scales is suggested mainly to create an operational tool for the practical biodiversity management in forestry and as a basis for forest policy; a complete long-term biodiversity strategy must take into account both interactions between the different geographical levels and the fact that different elements of biodiversity are dependent on different geographical scales, in different time perspectives.

The specialists involved in the BEAR project have agreed on a preliminary list of key factors, in principle relevant to all forests in Europe. It must be stressed that the great variation in ecological conditions and land-use in different parts of Europe as a rule, creates a variation in the relative importance of each key factor in different forest types. Taking this into account it has nevertheless been possible to present a common preliminary list of main key factors for European forests; cf. Table 4.

The key factors are very complex and the characteristics vary in different regions of Europe. The structural features of different European forest types varies; so do also the forestry practices and agriculture – cf. e.g. the differences as regards extensive grazing between the north and the south. Furthermore, forest landscapes differ in lowlands and mountains and forests in such a large area as Europe comprise different species compositions in different biogeographic regions. However, in principle this single list of key factors is considered to cover the variation in European forest biodiversity. Of course the relative importance of the key factors vary, as does the factors as such but this is dealt with in the step to follow later. It should be noted however that presently not all such data are being collected and in order for an assessment of forest biodiversity, a concerted data collection across Europe should be a priority.

Below are some further comments to the key factors in the table:

Table 4. A preliminary list of key factors of European forest biodiversity.

Scale	Structural key factors	Compositional key factors	Functional key factors
National/regional	Total area of forest with respect to: – Legal status/utilisation or protection – Forest ownership – Tree species and age – Old growth/Forest left for free development – Afforestation/deforestation	Native species Non-native or not "site original" tree species	FOR ALL SCALES: NATURAL DISTURBANCE: Fire Wind and snow Biological disturbance
Landscape	Number and type of habitats (incl. water courses) Continuity and connectivity of important habitats Fragmentation History of landscape use	Species with specific landscape-scale requirements Non-native or not "site original" tree species	HUMAN INFLUENCE: Forestry Agriculture and grazing Other land-use Pollution
Stand	Tree species Stand size Stand edge/shape Forest history Habitat type(s) Tree stand structural complexity Dead wood Litter	Species with specific stand type and scale requirements Biological soil condition	

Structural key factors

National/regional scale

TOTAL AREA OF FOREST: The total area of forests, their distribution, altitude and latitude are of importance for the biodiversity because a larger area will be more diverse as regard these factors and therefore also more diverse as regards habitats and habitat quality and consequently also regards biodiversity. For example, there exists a species-area curve, which means that one will find more species in a larger area than in a smaller one (e.g. Rosenzweig 1995).

AREA OF EXPLOITABLE FOREST (BY FORESTRY): Area of exploitable forest are of importance as the exploited forests normally have a lower number of species and another species composition than the non-exploited forests.

AFFORESTATION (YEARLY RATE): Afforestation is an effect of establishment of forests either by natural processes or by planting. Afforestation will heavily change the species composition at all scales (national/regional, landscape and stand). Open space species and species related to agricultural land will decrease and will be succeeded by forest species. In a region with low proportion of forests, also new species can colonize. After afforestation there will be a succession of species along the successional gradient, bush layer species will be succeeded by species related to more closed forests. Compare with reforestation, which means replacement of forest cover which has been removed either recently or in the past.

DEFORESTATION AND DESERTIFICATION (YEARLY RATE): Deforestation means removal of closed forest or open woodland, mainly for the reason to create land for agriculture. Similar to afforestation, the species composition is heavily influenced by deforestation. Forest living species will be succeeded by species related to open landscapes and agricultural landscapes. The term desertification has been used to describe the reduction in the biomass and productivity of the world's drylands that has become increasingly apparent over the past few decades (Calow 1998). Many areas that now have become more or less deserts were formerly capable of sustained agriculture and pastoralism. Naturally the process also creates a change in flora and fauna species composition utilizing these habitats. Zipper (1993) recognizes five identifiable patterns of deforestation in eastern United States – internal, indentation, cropping, fragmentation, and removal – and each has a distinct effect on habitat quality of forest patches.

FOREST PROTECTION (IUCN CATEGORIES): The area of protected forests will to some extent influence the possibility to maintain forest species at a national/regional level. The number and size of the protected forests, their connectivity and distribution will affect their effectiveness as refuge for forest species.

AREA OF OLD-GROWTH FORESTS: The term old-growth or ancient forests describes forests that have developed over a long period of time without experiencing catastrophic disturbance (Calow 1998). The age at which old-growth develops and the specific structural attributes that characterize old-growth vary with forest type, climate, site conditions and disturbance regime. Many species are related to old-growth forests, although relatively few are exclusively dependent on them. The old-growth forests have specific qualities which are of importance for different species groups as regards reproduction (e.g. nesting sites for birds, bark thickness, which influences substrate quality for wood living invertebrates), food searching, winter survival (e.g. food storing conditions for wintering birds), etc.

Landscape scale

FOREST COVER CONTINUITY: Forest cover continuity is of importance for biodiversity, especially for those species which are dependent on specific habitats and by more specialized species in the forests. Species with different dispersal capacity might have different possibilities to colonize and survive in forest stands, depending on forest cover continuity. At least three mechanisms can explain why species are dependent on long forest continuity 1) species which depend on a very long period of time to attain maximum size and to reproduce; 2) species which are associated with specific structural elements or processes present only in old-growth stands, for example very large trees and coarse woody debris (CWD); 3) species that are dependent on the specific microclimatic conditions in old-growth stands (Esseen et al. 1997).

FRAGMENTATION, MOSAIC AND ISOLATION OF FOREST HABITATS: A landscape perspective is required to understand how species are distributed across complex mosaics of habitat patches (Dunning et al. 1992). The ability of organisms to migrate and disperse across the forest landscape depends to some degree on the spatial structure of the landscape. Therefore, fragmentation is a central issue in the conservation and management work. The term fragmentation is used in conservation biology to refer to the process that converts large areas of relatively uniform vegetation into a mosaic of small patches of vegetation of different age classes and wildlife habitat potential across the landscape (Kimmins 1997a). When a forested landscape becomes fragmented, there occur changes in the physical environment of forests. Fragmentation influences the biodiversity by reduction of habitats, but there will also be smaller sizes of the habitat fragments left and increased isolation of the fragments. The consequences will be smaller habitats for forest species, which affects population sizes in each habitat fragment, and increased risk for extinction of isolated populations among species with small dispersal capacity. In fragmented systems the major impacts on vegetation fragments often arise from changes in the surrounding matrix. For example, hydrological changes in the surrounding landscape (e.g. rising or falling water tables) can have dramatic impact on remaining patches of native vegetation (Hobbs 1999). Actually, fragmentation and loss of natural habitat are regarded as major threats to biological diversity (Wilcove et al. 1986, Dempster 1991, Wiens 1995). Yet, when considering the entire dimension of deforestation, fragmentation is only one part of the process (Zipper 1993). Species differ in their responses to fragmentation. As a result, the loss of species with a reduction in area and changes in habitat is nonrandom (Wiens 1990). However, the effects of loss of habitat are non-linearly related to the loss of habitat. Fragmentation leads to shifts in community composition and structure that are not evident in simple species-area formulations (Wiens 1990, Andrén 1997). There are several evidences for thresholds in the proportion of suitable habitat in the landscape that are important for changes in biodiversity, and from a conservation point of view, it is important to find these potential thresholds. From a conservation point of view there should be certain proportions of different habitat types in the landscape, which needs plans on landscape level (Andrén 1997).

FOREST CONNECTIVITY: Connectivity is a feature of landscape mosaics and could be measured as the degree to which patches of a given type are joined by corridors into a lattice of nodes and links (Wiens et al. 1993). It could also be expressed as the degree to which isolation is prevented by landscape elements which allow organisms to move among patches (Calow 1998). Thus, connectivity influences the capacity of species to utilize different parts of a fragmented landscape. As the landscape becomes divided into smaller parcels of habitat, landscape connectivity (i.e. the functional linkage among habitat patches) may suddenly become disrupted, which may have important consequences for the distribution and persistence of populations (White and Crist 1995). Landscape connectivity depends on the abundance and spatial distribution of habitats, but may also be influenced by habitat specificity and dispersal abilities of species utilizing the landscape. For example, habitat specialists with limited dispersal capabilities presumably have a lower threshold to habitat fragmentation than vagrant species (White and Crist 1995).

ECOSYSTEM DIVERSITY: The term diversity normally refers to the variety and abundance of species at a specified place and time. Consequently, ecosystem diversity could mean the variety and abundance of ecosystems, but should also, like biodiversity, include the dimension of evenness, i.e. a measure of how equally abundant the different ecosystems are.

WATER COURSES: Water courses in the forested environment increase the ecosystem diversity and thus also the species biodiversity. The riparian zone along water courses is the river band or the strip of land that separates the land from the streams, lakes, etc. Trees and shrubs growing there are necessary to provide an input of CWD to the water (e.g. Samuelsson et al. 1994). The CWD in waters is important to create habitats for fish and aquatic invertebrates and they are strongly influenced by channel structures created by the wood, as well as the sediments of organic and inorganic matters it storage (e.g. Harmon et al. 1986). Although it seems to be the case that abundance, diversity and degree of social organization of insect taxa associated with wood are much lower than in terrestrial ecosystems, the wood is still important even for insects, because when CWD is abundant, a specialized fauna has evolved that is closely associated with woody debris (Harmon et al. 1986).

DIRECT HUMAN IMPACT: The biodiversity in forested landscapes is heavily influenced by human activities, such as forest management systems (which influence the fragmentation of the forests, tree species composition, tree age distribution, etc.) forest road network (which might influence biodiversity by disturbance of rare species, and by influencing dispersal patterns), forest grazing by cattle, hunting regulations by such herbivores as moose, fire protection, construction of buildings, etc.

Stand scale

SIZE: The size of a stand is of importance for the biodiversity because there normally is a direct relation between the size of a forest stand and the number of inhabiting species, the so-called species-area relationship, which characterizes the diversity of all taxa and is known from all biomes all over the world. In a managed forest, forest stands create patches in fragmented landscape. Patch size is of major ecological importance; for example, large patches contain interior species absent in patches of small size (Forman 1995).

SHAPE: The shape of a stand is of importance for biodiversity because it influences the boundary types and the relation between the edge length and area and the core area of a stand. Shape refers to the form of a area (two-dimensional), as determined by variation in its margin or border (Forman 1995). For example, an elongated patch is less effective in conserving internal resources than a circular patch. This is considered to be true for protecting interior species, as well as for species requiring a distance from the effect of human activities (Forman 1995).

TREE STAND STRUCTURAL COMPLEXITY: The structural component of the tree stand determines to some extent the biodiversity by allowing species utilizing different niches in the stand to co-exist. Not only the tree species composition and age structure influence the diversity, but also the vertical and horizontal patterns in general. Vertical structure is influenced by components as very old trees, trees with heavy load of epiphytic lichens, broken top, stag-headed and leaning trees, trees with holes and cavities, dead standing trees (snags), fire-scarred trees, snags and stumps, stumps with uneven surfaces and large-sized logs in various

stage of decomposition. The horizontal structure is influenced by a patchy distribution of tree species and tree stems, uneven-aged stand structure and multi-layered tree canopies, but also by such processes as post-fire succession, succession with tree-species replacement, self-thinning and gap formation (Esseen et al. 1992). Raivio (1988) found that the most important factor determining bird densities was the luxuriance of the habitat, but the bird communities were also affected by tree species composition and the size class distribution of the trees.

FOREST REGENERATION: The processes of regeneration vary greatly between communities. Quite often regeneration occurs naturally after large-scale disturbances (such as fire, wind or flooding). In other cases regeneration occurs after a small-scale disturbance caused by a fall of single trees creating gaps in the forest. Regeneration from seeds will be influenced by climate, seed predation by seed-eating animals, grazing activities, etc.

DEAD WOOD QUALITY AND AMOUNT. LITTER (QUALITY AND AMOUNT): A large proportion of organisms living in the forests are dependent on the presence of dead wood (not least beetles and lichens). Many birds are dependent on dead wood for food-searching and nesting. Also the litter produced after for example clearcutting are utilized by invertebrates, and thus the quality, e.g. the dimension of branches, and the amount of branches left is of importance for biodiversity. Logs of Norway spruce *Picea abies*, aspen *Populus tremula* and natural stumps, chimney stumps, snags, pollards and other old trees, are defined as key elements in woodland habitats in the Swedish forests by Nitare and Norén (1992). Jonsell et al. (1998) have evaluated which qualities of dead wood have the highest conservational value for invertebrates by recording the substrate requirements for all 542 saproxylic (wood-living) red-listed invertebrates in Sweden. They conclude that we need a diversity of substrate types and management methods to maintain all saproxylic species. Some 59% of the invertebrate species can live in sun-exposed sites, but some species, especially those living in the last successional stages, are dependent on shaded sites. Hollow tree trunks are important micro-habitat. Sixty-four of the 107 species living there are specialists. For a review of dying and dead trees and their importance for biodiversity, se also for example Samuelsson et al. (1994).

WATER COURSES: Similar to the situation at a landscape level, water courses in forest stands increase the structural complexity, and thus also the biodiversity. Not least the vegetation along a water course, creating corridors, is of importance. River corridors are by far the most dynamic place in many landscapes. The stream corridor is exceptionally diverse environmentally, and hence normally supports a high species richness, sometimes the highest in the landscape (Forman 1995).

Compositional key factors

National/regional scale

NATIVE SPECIES: The prevailing native tree species composition in a country or region generally determines the plant and animal species composition. If the dominant trees in a region, for example, are broad-leaved, the biodiversity is quite different from a mixed tree species forest or a coniferous forest.

ALIEN SPECIES: If the native species are succeeded by alien species, the species biodiversity connected to the alien species are normally lower, because the time needed for example for invertebrates to adjust to the new species. The time needed for this process is to a large extent dependent on the area covered by the alien species, but is also dependent on how closely related to the prevailing native species the introduced species is.

Landscape scale

SPECIES WITH LARGE AREA REQUIREMENTS: Some animal species are dependent on large home ranges and thus they do not occur in, for example, heavily fragmented landscapes.

SIMILARITY TO POTENTIAL NATURAL VEGETATION: The similarity to potential natural vegetation could be a factor of importance when evaluating the potential value of a landscape from a biodiversity point of view, if the aim is to maintain the natural biodiversity in the landscape.

Stand scale

STAND TYPE SPECIFIC SPECIES: Some species require quite specific habitat conditions to survive. It could be, for example, specific moist conditions, which is essential for many lichens, and also particular light conditions specific both for plant and animal species. These particular conditions are to some extent dependent on for example stand scale and configuration.

BIOLOGICAL SOIL CONDITION: Even the soil conditions in a forest stand is to some extent dependent on light and moisture, but also on the tree species composition. For example litter from deciduous trees can heavily influence the soil conditions in a forest stand, which also will influence the species composition utilizing the soil as habitat.

ALIEN SPECIES: Alien species among trees, for example, influence the biodiversity of a forest stand as such, but they may also change the biodiversity indirectly by influencing the species composition of insect and plant species utilizing the alien trees as food and substrate, and also by changing the food abundance for insectivorous bird species. In general, in cases where an exotic species is introduced it may become successful, and thereby influence the native

biological diversity if 1) it has a greater competitive ability than native species, 2) through expansion, it can modify the ecosystem to its own advantage and thereby increase its population, or 3), disturbance regimes are changed in a way that favor the introduced species. These three factors vary to a smaller or larger degree over time (Huston 1994). There often exists a relation between disturbance of ecosystems and introduced species. Introduced species are common in disturbance corridors, for example in natural ones like riparian corridors, or disturbed areas with anthropogenic origin. Such disturbed areas are roads, trails, pipelines, powerlines, livestock grazing, and clearing for agriculture or timber harvesting (D'Antonio et al. 1999). Non-indigenous species are frequently spread from such corridors to adjacent undisturbed habitats. However, whether or not invasion of adjacent habitats really occurs is likely a function of the habitat type, the biology of the invader, and the length of time since invaders entered the area (D'Antonio et al. 1999). Introduced species can affect the size, frequency, intensity, or type of disturbance by being agents of disturbance themselves or fueling disturbances, but they can also alter the response of the community to disturbance (D'Antonio et al. 1999).

Natural disturbance/management key factors

Natural disturbance

One of the more commonly accepted definitions of disturbance in the ecological context is: "A disturbance is any relatively discrete event in time that disrupts ecosystem, community or population structure and changes resources, substrate availability or the physical environment" (Pickett and White 1985). According to the terminology by Oliver and Larson (1990), disturbances can be divided into two main groups, "major" and "minor". A major disturbance is defined as one that kills practically all the trees above the forest floor (scale of stand), whereas a minor disturbance leave groups of trees alive (Engelmark 1999). Fires generally create large openings in the forested landscape, although there is a great variety in boreal fire regimes and related ecological responses (e.g. Bergeron et al. 1997, Engelmark 1999). In areas where fires, extensive insect outbreaks, or other major canopy disturbances are unimportant, minor openings in the canopy play a central role for ecosystem rejuvenation (Kuuluvainen 1994, Engelmark 1999, see also Platt and Strong 1989, Denslow and Spies 1990). Such gap disturbances, caused by for example falling trees, also appear to be common (e.g. Kuuluvainen 1994). Disturbances may be characterized by their frequency (mean number of events per time period), intensity (physical force of the event per unit area per unit time), and extent (area disturbed) within a domain of interest (Pickett and White 1985). The timing of the disturbance is an important factor determining the ecological effects of, for example, a forest fire.

FIRE: Fire is an important disturbance factor in both boreal (e.g. Zackrisson 1977, Bergeron et al. 1998, Engelmark 1999) and Mediterranean ecosystems (e.g. Naveh 1974, Faraco et al. 1993, Moreno et al. 1998, Pérez and Moreno 1998), although its role may have been overemphasized in boreal systems (Bergeron et al. 1998). However, the important role of fire in the landscape can be difficult to evaluate because fire sometimes has operated in close combination with uncontrolled grazing and other human interference (Naveh 1974). Effects of fire will depend on the site type and the intense of fire and how often a fire occurs. This has been used in the boreal region for developing a model for forest management regimes (Angelstam 1998a, b). Normally fires occur with relatively low intensity and high frequency in pine forests. In Norway spruce stands, the dominating forests on mesic sites, fires are more intense but less frequent. The ecological consequences of a fire will be exposed mineral soil, at least at severe fires, nutrient release, tree death (except for some thick-barked trees, primarily pines), and natural regeneration as start of forest succession. Also the structure of the stands will change. For example, in pine forests some old pines survive a fire, with fire-scarred, living and dead trees standing as result, while others die, with snags and downed logs as a result, which creates an uneven-aged, multi-layered, and all-sized stand structure and patchy distribution of trees. The fraction of deciduous trees will increase (Esseen et al. 1997, Fries et al. 1997).

WIND AND SNOW: Sometimes disturbances are created by wind and snow, with broken, stag-headed, leaning, dead and dying trees as result, as well as exposure of the soil surface, uprooting of trees, and creating sunlit open areas. Natural regeneration and succession processes start. The ecological effects of wind disturbance are dependent on magnitude, intensity and severity. The return time, i.e. the frequency of disturbance events over time, is the key to understanding forest turnover times and dynamics at the landscape scale. The disturbance regime is highly variable in space and in time, but a full understanding of windthrows as ecological disturbance events is difficult to get in, for example, temperate forests, due to loss of intact forests as an effect of agriculture and timber harvesting. On a small scale wind and snow create gap dynamic processes, which change the structure of the stands as single trees are falling, and trees are broken, creating uprooting, exposure of mineral soil, resulting in regeneration of down logs, stumps and tussocks in the sunlit openings in the forest stands. When trees are wind-thrown, a pulse of new resources becomes available, with increased light, moisture, and nutrients. Microsites on the forest floor (wind-throw mounds, pits, stumps, and logs) provide new substrates in a matrix of soil covered with litter. The composition of the trees in the gap will change with an uneven age distribution and uneven dimensions of the trees. In conifer forests the proportion of deciduous trees increase (Esseen et al.

1992, Kuuluvainen 1994). In many areas the tree-line is wind-determined, and at its upper altitudinal limit, the trees are heavily influenced by wind. The trees grow in the lee of boulders and create forest patches oriented in the direction of the prevailing wind (Komarkova and Wielgolaski 1999).

BIOLOGICAL DISTURBANCE: Browsing and grazing by large herbivores like roe deer, red deer, moose and reindeer, and tree defoliation by herbivorous insects like the autumnal moth *Epirrita autumnata*, are examples of disturbances structuring forests (e.g. Engelmark 1999). During outbreaks *Epirrita autumnata* causes extensive mountain birch mortality at the treeline in Fennoscandia (Haukioja et al. 1988). The beaver causes such changes in structure in the forested landscape that it could be regarded as a keystone species for many species living in the beaver dams (e.g. Naiman et al. 1986, Johnston et al. 1993). Grazing or browsing can be regarded as a disturbance acting more or less continually in forest, and may therefore be termed "chronic", in contrast to "pulsed" disturbances such as fire (Englemark 1999). Although it could be discussed whether or not insect herbivores should be regarded as agents of disturbance, rather than simply as ecosystem components that respond to environmental change, herbivore outbreaks can dramatically alter ecosystem structure and function in much the same way as other disturbances (Schowalter and Lowman 1999). Herbivory can affect forests over a wide range of spatial scales. In some cases, a single tree may be targeted, in other cases herbivory can affect thousands of square kilometers (Schowalter and Lowman 1999). Herbivory may be considered as a keystone ecological process in an ecosystem, which can serve as an indicator of disturbance (Lowman 1999).

Human influence

Fire and grazing are important factors in the landscapes influenced by man. As expressed by Naveh (1974), for example, it is obvious that fire has acted, not as a wholly destructive force, but as a powerful selective and regulatory agent throughout geological, biological and cultural evolution of the Mediterranean landscape. The role of livestock in the change of the Mediterranean forests have been stressed by, for example, Thirwood (1981), who considers grazing by domestic animals among the major causes of forest degradation. Tsoumis (1985) even regards pastoral economy as a more important factor for deforestation than agricultural clearances.

SILVICULTURE: By silviculture activities aiming at increasing the production, the forestry has heavily influenced the prerequisite for maintaining the kind of biodiversity existing in natural forests in the region by changing both processes, structures and composition of the original forests. Not least the prevailing silviculture methods used up till recent time have influenced the structure of present-day forests, and thereby also the structure of the forests, with consequences for biodiversity. As example, the coppice system could be mentioned, as well as the single tree selection system. The short rotation length of the coppice system creates habitats suitable for birds characteristic of the early open stages of the forest succession. Species richness declined with coppice age, particularly after canopy closure (Fuller and Moreton 1988, se also Avery and Leslie 1990, Fuller and Henderson 1992, Sage and Robertson 1996). Atlegrim and Sjöberg (1996) demonstrated that in uneven-aged boreal *Picea abies* forests in north-eastern Sweden, herbivorous insect larvae feeding on bilberry *Vaccinium myrtillus* were influenced by clear-cutting and single-tree selection. Although standing dead trees and fallen logs are essential to many organisms and biological processes within forest ecosystems, such structures have rarely been retained within managed forests (Franklin 1988). However, in present-day forestry, for example considerations aiming at maintaining biodiversity by trying to mimic natural disturbance processes and by keeping structural components in the forests of importance for plant and animal species, are implemented in forestry policies and management plans in many countries (e.g. Sjöberg and Lennartsson 1995). Silvicultural treatment also influences disturbance patterns and processes in the forests. For example, when trees are thinned, remaining trees often succumb to high winds in their newly opened canopy. Increased numbers of wind-throws are documented in thinned stands, especially where topographic risk factors are high, and where thinned stands are also fertilized.

SPECIAL FORESTRY TREATMENT: In some areas forests are treated with specific care or they are not used for forestry because they have a function as protection forests around water resources, or they are used, in the Alpine region for example to protect from avalanches. Another purpose could be as protection from erosion. In such forests the harvesting aspects are often of minor importance. For that reason, such forests more or less act as forest reserves, and thereby they are important also from a biodiversity point of view, as the structures and processes in such forests are more similar to the situation in original forests of the region.

AGRICULTURE AND GRAZING: A direct effect of agriculture on the forest ecosystems is of course the reduction in forest area, but in older times agriculture activities also heavily influenced the forest ecosystems as such by a variety of multiple use of the resources. Even if such an activity as for example slash and burn management has disappeared since a long time ago, the ecological long lasting effects to some extent are influencing the forests even today, e.g. by the composition of tree species.

Before the forests had got their present economic values for wood production, they were normally used for grazing by cattle, goats and sheep. In some areas the forests were burned to increase the growth of valuable plant species. This has been the case in Fennoscandia, and is still to some extent the case in southern Europe. However, the Mediterranean ecosystems were not affected only by grazing per se but by the whole livestock economy, which involved other activities, (Papanastasis 1998). Shepherding, movement (nomadism and transhumance) sheltering, milking, cheese making and, especially, occupational burning, are such examples. According to Papanastasis (1998), the latter activity was instrumental in shaping Mediterranean ecosystems more than grazing, because fire was widely used not only to open grazing land but also to suppress unwanted, chiefly woody, vegetation in rangelands and is still used in several parts of the Mediterranean region.

Although overgrazing still is regarded as a problem in the Mediterranean area, the opposite could be a problem too. For example Rackham and Moodey (1996) pointed out that disappearance of grazing-prone species or the piling up of flammable biomass which often leads to devastating wildfires. According to Papanastasis (1998) it is widely accepted nowadays that the extensive wildfires that occur in the southern European countries over the last few decades are caused by the drastic reduction of the human activities in the region, including livestock grazing due to rural immigration to urban centers.

URBAN FORESTS: For many people it is important to have access to a nearby forest for recreation activities (Hörnsten and Fredman 2000), and many municipalities have the main responsibility for management of recreation interest in urban settings. An important part of local planning is formed by physical planning and management – in which outdoor recreation possibilities have to be taken into account (Jensen 1995). That also includes planning of forests in urban settings. As regards the areas which are set aside for nature protection and/or out-door recreation, the main problem will be to maintain the various nature and environment qualities simultaneously with the improvement of facilities for different users. This is caused by a variety of whishes and demands from different user groups which cannot be met simultaneously everywhere (Jensen 1995). The management of urban forests can be managed in such a way that the variation in age structure, tree species composition etc., is maintained, and thereby also biological diversity.

POLLUTION: Pollution, mainly from the air, will influence forest ecosystems in many ways. By acid rain, for example, the pH level in the soil will decrease, which will influence the composition of the soil organism community, cause thinning of needles in conifers, and at extreme conditions even kill trees. By nitrogen deposition the forests will be fertilized and thereby cause a shift in plant species composition directly, as some species react positively as an effect of higher nitrogen concentration, while other react negatively, but also indirectly by changing the competition processes in the plant communities. Subsequently this will influence also for example the insect species related to the different plant species. Climate change, as effects of pollution, are supposed to have the potential to strongly affect the boreal ecosystems. Fire and herbivory have since long time been identified as important key disturbances in that region. However, the effects of climate change on these disturbances, and consequently on the ecosystems, might even be greater than the ecosystem changes due to climate change per se (Davis and Botkin 1985, Fleming 1996, se also Hofgaard et al. 1999).

EROSION: There are few quantitative data concerning erosion rates and the extent of erosion on agriculture and forestry (Pimentel and Harvey 1999). However, it has been estimated that ca 80% of the world's agricultural land suffers from moderate to severe erosion (Oldeman et al. 1990, Pimentel 1993, Speth 1994). By diminishing soil organic matter and overall soil quality, erosion reduces biomass productivity in ecosystems. At the same time, biodiversity is significantly reduced (Pimentel and Harvey 1999).

In Chapter 4 more details as regards special characteristics of key factors in different forest types are given.

3.2 The concept of forest types for biodiversity assessment

In Chapter 3.1 the concept of key factors of forest biodiversity is introduced and it is stated that in spite of the fact that it has been possible to agree on a single preliminary list, the relative importance of key factors varies greatly in different European forests, as do the factors as such.

A strategy for assessment and management of European forest biodiversity must consider this variation of key factors of biodiversity. There is no single specific model suitable for all European forests. The BEAR specialists suggest that biodiversity assessment and management should be based upon a number of forest types; each being reasonably homogenous as regards the key factors of forest biodiversity. The variation in the key factors is the criteria for identifying the BEAR European scheme of "Forest Types for Biodiversity Assessment FTBA".

The preliminary scheme of Forest Types for Biodiversity Assessment FTBAs is thus established in order to be suited for the management of forest biodiversity and to meet with the requests of:
• European level harmonisation: To support the process of protection of forests in Europe and other pan-European initiatives as regards biological and landscape diversity, the BEAR scheme of Forest Types for Biodiversity Assessment FTBAs must, at some level, be feasible for assessing biodiversity in Europe and also for each European State.

• Practical feasibility on forest management unit level: The Forest Types for Biodiversity Assessment FTBAs must be acceptable for the managers of forest biodiversity, at landscape and smaller management unit level (i.e. single forest stands). This puts several requests on the FTBAs: e.g. each FTBA must correspond to or be easily recognised in terms of types already used in national forest type schemes.

The units of a scheme of Forest Types for Biodiversity Assessment FTBAs should be those forest types we need to distinguish because they need different considerations in the planning of how to manage forest biodiversity. As this takes place at different scales (national, regional/landscape and stand level) the FTBA scheme must be relevant for all these levels. The scheme and it forest types must be hierarchical, presenting units relevant to different scales.

In a preliminary scheme the BEAR experts agreed on 33 Forest Types for Biodiversity Assessment FTBAs of which:
• FTBA 1–24 reflect certain types of potential natural vegetation on different ecological sites;
• FTBA 25 is the laurel forest, nowadays mainly confined to the Macaronesian region
• FTBA 26–27 are examples of purely antropogenic, albeit from biodiversity point of view important forest types developed by "traditional land-use";
• FTBA 28–33 are plantations, cf. Appendix 6. As regards FTBA 33 specify species (e.g. Douglas fir, Sitka spruce).

Table 5 below shows the preliminary list of common Forest Types for Biodiversity Assessment FTBAs with notification of key factor(s) – at stand level – decisive for the identification of each FTBA.

The relative importance of the key factors assigned to each FTBA in Table 5 are "subjective expert opinions" by the BEAR partners based upon current knowledge. The intention of Table 5 is to provide a general overview how the FTBAs have been identified. In Chapter 4 the key factors of particular importance for each FTBA is presented in some more detail.

Furthermore for European level harmonisation the BEAR experts found the FTBAs rather well to correspond to relevant units of the Map of Potential Natural Vegetation of Europe (Bohn 1994, 1995, cf. Appendix 4). On a European level types of Potential Natural Vegetation and Biogeographic regions the FTBAs are distributed as shown in Table 6.

From the comparison between CORINE land cover nomenclature for Europe and the main formations based on Potential Natural Vegetation, it can be concluded that rough correspondences exist. This has made it possible to make reference to the CORINE system in the descriptions of each European-level Forest Type for Biodiversity Assessment (Chapter 4).

As expected from its origin, CORINE classification

Table 5. A presentation of the Forest Types for Biodiversity Assessment FTBAs with notification of relative importance of key factors of biodiversity at stand scale. The criteria for the identification of each FTBA is uniqueness as regards importance of most relevant the key factors.

Forest Type for Biodiversity Assessment FTBA	Relative importance of stand scale key factors of forest biodiversity: 0 = none, 1 = slight, 2 = moderate, 3 = major																
	Structural: S1 Tree species S2 Stand size S3 Edge characteristics S4 Forest history S5 Habitat type(s) S6 Tree stand structural complexity S7 Dead wood S8 Litter								Compositional: C1 Species with specific stand type and scale requirements C2 Biol. soil condition		Functional: Natural disturbances: N1 Fire N2 Wind and snow N3 Biological disturbance (incl. pests)			Functional: Human influences H1 Forestry H2 Agriculture and grazing H3 Other land-use H4 Pollution			
	S1	S2	S3	S4	S5	S6	S7	S8	C1	C2	N1	N2	N3	H1	H2	H3	H4
1. Subalpine conifer vegetation in nemoral zone	3	1	2	3	3	3	3	2	2	2	1	3	1	1	3	1	2
2. North boreal spruce forest	3	3	3	3	3	3	3	3	3	2	3	3	2	3	1	1	2
3. North boreal pine forest	3	3	3	3	3	3	3	3	3	2	3	3	2	3	1	1	2
4. Middle boreal spruce forest	3	3	3	3	3	3	3	3	3	2	3	3	2	3	1	1	2
5. Middle and south boreal and hemi-boreal pine forest	3	3	3	3	3	3	3	3	3	2	3	3	2	3	1	1	2
6. South boreal forest	3	3	3	3	3	3	3	3	3	2	3	3	2	3	1	1	3
7. Hemiboreal spruce and fir-spruce forests	3	3	3	3	3	3	3	2	3	2	3	3	2	3	1	2	3
8. Mixed spruce and fir forest	3	1	2	3	3	3	3	2	3	2	1	2	2	3	2	1	3
9. Mixed oak forest	3	3	2	3	3	3	3	2	3	2	1	1	1	3	0	1	2
10. Ashwood	3	3	2	3	3	3	3	2	3	2	0	2	1	3	2	1	?
11. Mixed oak-hornbeam forest	3	1	2	3	3	3	3	2	3	2	0	1	1	3	2	1	2
12. Lowland and submontane beech forest	3	1	2	3	3	3	3	2	3	3	0	1	0	3	2	1	2
13. Montane beech and mixed beech-fir-spruce forest	3	3	3	3	3	3	3	3	3	3	1	2	1	3	2	0–1	0–1
14. Mediterranean and Submediterranean mixed oak forest	3	3	3	3	3	3	3	3	3	3	1–2	1	1	3	2–3	1	0–1
15. Mediterranean broad-leaved sclerophyllous forests and shrub	3	2	2–3	3	3	2–3	3	3	3	3	3	0–1	0–1	3	3	2	0–1
16. Mediterranean and Macaronesian coniferous forests, woodlands	3	3	2–3	3	3	3	3	3	3	3	3	1–2	1–2	3	2–3	1–2	0–1

Table 5. Cont.

Forest Type for Biodiversity Assessment FTBA	Relative importance of stand scale key factors of forest biodiversity: 0 = none, 1 = slight, 2 = moderate, 3 = major																
	Structural: S1 Tree species S2 Stand size S3 Edge characteristics S4 Forest history S5 Habitat type(s) S6 Tree stand structural complexity S7 Dead wood S8 Litter								Compositional: C1 Species with specific stand type and scale requirements C2 Biol. soil condition		Functional: Natural disturbances: N1 Fire N2 Wind and snow N3 Biological disturbance (incl. pests)			Functional: Human influences H1 Forestry H2 Agriculture and grazing H3 Other land-use H4 Pollution			
	S1	S2	S3	S4	S5	S6	S7	S8	C1	C2	N1	N2	N3	H1	H2	H3	H4
17. Atlantic dune forest	3	3	2	2	2	2	3	3	3	3	2	3	2	1	2	3	?
18. Ombrotrophic mires	3	3	0	3	3	0	0	0	3	0	2	1	1	1	1	1	3
19. Arctic-subarctic mires	3	?	2	?	3	?	?	?	3	?	0	3	3	0	1	1	3
20. Minerotrophic mires incl. swamp forest	3	3	0	3	3	3	3	0	3	0	0	0		2	3	1	3
21. Swamp and fen forests, alder	3	3	0	3	2	2	2	0	3	0	0	0	0	2	3	1	3
22. Swamp and fen forests, birch	3	3	0	3	2	2	2	3	3	0	0	0	0	2	3	1	3
23. Flood plain (alluvial and riverine) forests	3	3	2	3	3	2–3	3	1	1–2	?	1	3	1	3	1	2	?
24. Mediterranean and Macaronesian riverine woodlands and gallery forests	3	3	3	3	3	3	3	2	3	2	2	3	2	3	3	3	3
25. Laurel forest	3	3	?	3	3	3	3	3	3	2?	?	?	?	1	?	?	?
26. Hedgerow	3	1	1	3	3	3	0	0	2	0	0	2	0	0	2	0	0
27. Chestnut coppice	2	2	1	3	2	3	3	1	3	2	3	3	3	2	3	2	2
28. Pine plantation	3	3	2	3	3?	2	3	2	2?	?	1	3	3	3	2	1–2	?
29. Spruce plantation	3	3	2	3	3?	2	3	2	2?	?	1	3	3	3	2	1–2	?
30. Poplar plantation	3	0	0	2	2	0	0	3	2?	3	1	1	1	3	0	0	0
31. *Robinia* plantation	3	0	0	0	0	0	0	1	?	1	1	0	0	3	0	0	0
32. *Eucalyptus* plantation	3	0	0	0	0	0	0	0	?	0	1	0	0	3	0	0	0
33. Other plantation	3	0	0	0	0	0	0	0	2	0	2	0	0	3	0	0	0

derived from satellite images, clusters, in level 3 situations with similar physiognomy regardless of the species composition, while Bohn's classification tends to cluster primarily based on climatic characteristics resulting in more understandable geographical patterns of tree species distributions. Therefore Bohn's classification was selected as basis to define the FTBAs whereas CORINE classification is suggested as more operational for use in the BET System. For this it is suggested that, for the European scale, the level 3 of CORINE classification should be used, while for national and landscape scales it is appropriate to supplement this with further subdivisions.

Table 6. The Forest Types for Biodiversity Assessment FTBAs, their corresponding Potential Natural Vegetation Type(s) and occurrence in European Biogeographic Regions (cf. Fig. 3).

Forest Types for Biodiversity Assessment FTBA	Corresponding Potential natural vegetation type(s) Main formation, first division *)	Occurs in the following Biogeographic Region (Council Directive 92/43/EEC)					
		Boreal	Atlantic	Conti-nental	Alpine A=Alps P=Pyrenees C=Carpathian S=Scandinavian Alps	Mediter-ranean	Macaro-nesian
1. Subalpine conifer vegetation in nemoral zone	C3				A, P, C		
2. North boreal spruce forest	D1	•			S		
3. North boreal pine forest	D10	•			S		
4. Middle boreal spruce forest	D2	•			S		
5. Middle and south boreal and hemiboreal pine forest	D11	•			S		
6. South boreal forest	D3	•					
7. Hemiboreal spruce and fir-spruce forests	D8 + D12	•		(•)			
8. Mixed spruce and fir forest	D9			•	A, C		
9. Mixed oak forest	F1	•	•	•			
10. Ashwood	F2		•				
11. Mixed oak-hornbeam forest	F3		•	•			
12. Lowland and submontane beech forest	F5 a		•	•	A, P, C		
13. Montane beech and mixed beech-fir-spruce forest	F5 b			•	A, P, C	•	
14. Mediterranean and Submediterranean mixed oak forest	G2+G3		•	(•)		•	
15. Mediterranean broad-leaved sclerophyllous forests and shrub	J1-J8					•	•
16. Mediterranean and Macaronesian coniferous forests, woodlands	K1-K4					•	•
17. Atlantic dune forest	P1		•				
18. Ombrotrophic mires	S1	•	•				
19. Arctic-subarctic mires	S2				(S)		
20. Minerotrophic mires incl. swamp forest	S3	•					
21. Swamp and fen forests, alder	T1	•	•	•	(A)		
22. Swamp and fen forests, birch	T2	(•)	•				
23. Flood plain (alluvial and riverine) forests	U1+U2		•	•			
24. Mediterranean and Macaronesian riverine woodlands and gallery forests	U4					•	•
25. Laurel forest	–						•
26. Hedgerow	Traditional		•	(•)		•	•
27. Chestnut coppice	Traditional		(•)		(A)(P)(C)	(•)	
28. Pine plantation	Plantation	•	•	•		•	•
29. Spruce plantation	Plantation	•	•	•			
30. Poplar plantation	Plantation		•	•			
31. *Robinia* plantation	Plantation			•			
32. *Eucalyptus* plantation	Plantation					•	•
33. Other plantation	Plantation		•			•	

• = of major importance in the biogeographic region (•) = of minor importance in the biogeographic region
*) See Appendix 4.

Chapter 4.

Major European forest types for biodiversity assessment

Here the 33 identified Forest Types for Biodiversity Assessment FTBAs relevant for European and national level assessment and management of forest biodiversity are presented in more detail. Each presentation contains:

1. Name of FTBA.
2. Corresponding types of Potential Natural Vegetation (cf. Table 6 and Appendix 4).
3. A map showing the potential natural distribution of the FTBA (based upon the Map of Potential Natural Vegetation of Europe 1:10 mill. (Bohn 1994, 1995). Restricted areas of distribution of the FTBA may not be indicated on the map.
4. A short description of dominant tree species and other major elements of biodiversity are presented as well as the characteristic site conditions. The most relevant key factors of forest biodiversity, cf. Chapter 3.1, are de-

scribed in some detail. Emphasis is given to disturbance, land-use regimes and succession patterns.
5. Corresponding types according to national schemes currently in use.

As forest type schemes, as well as vegetation classifications systems, are already widely used by relevant EU and Governmental agencies, planners and forest managers it is a prerequisite for the practical feasibility that the Scheme of Forest Types for Biodiversity Assessment does not differ too much from those systems already in use or that it is clearly shown how the units correspond to those of existing schemes. In the more detailed description of each Biodiversity Forest Type below a reference is made to corresponding practically used forest types in a number of European countries, cf. Appendix 5.

Subalpine conifer vegetation in nemoral zone C3

Characteristics, ecological conditions and main tree species

This forest type does not have a widespread spatial distribution in Europe but under specific climatic conditions it may cover an important altitudinal belt in all Alpine regions through central Europe and the Mediterranean area from Spain to Greece and even as far as the Caucasian range. It covers a well defined but sometimes very narrow altitudinal belt. Therefore, this forest type is not well represented on the European map because of its disjunctive and sometimes very small scaled distributional areas. In most cases it cannot be described on the chosen scale of the map. This altitudinal belt had originally a larger range but due to human activities over thousands of years it lost the largest part of its original spatial distribution to pasturing dependent alpine vegetation such as alpine meadows.

Four main areas of distribution exist but in the following ecological description we will be focusing on the specific alpine situation:

- Pine dominated (*Pinus uncinata* and *Pinus sylvestris* var. *iberica*) open and light-flooded forests form the subalpine forest vegetation of the Pyrenees and the oromediterranean altitudinal vegetation belt of the Iberian peninsula.
- In the Alps the subalpine coniferous vegetation covers a small altitudinal belt of open-structured mixed *Pinus cembra-Picea abies-Larix decidua*-forests, partly dominated by larch (especially on limestone sites of the southern and northern Alps).
- The third isolated part of the area in which this forest type is to be found is located in the Carpathian mountains and at the Balkan range, especially in the Greek mountains.
- No closer consideration of the forest type occurring in the Caucasian range.

This forest ecosystem forms the highest upper timberline in the Alps (1700–2500 m altitude) especially in the inner part, depending on the magnitude of the mountain massive. The prevalent parent material is silicate, which favours *Pinus cembra*. In the peripheral alpine region where limestone is prevalent, the timberline is lower. *Pinus cembra* islands are therefore rare and larch forests more common. The precipitation varies from 800 to 1800 mm depending on the geographical location, with greater rainfall in the western Alps and less in the inner and eastern Alps. On siliceous substrates, podzolic soil types prevail, and on limestone, rendzinas are dominant. Both are characterised by a thick layer of raw humus. Climatic, site and topographic conditions are very harsh, with a growing season of only three to four months. In the mountainous zone of some prealpine mountains (e.g. Vosges, Black forest, Hercynic mountains, Bayrischer Wald) *Pinus mugo* subspecies may actually form the timberline vegetation. Most of these high montainous to subalpine "Krummholz" shrub forests are not natural. They represent succession stages after pasture, derived from the native peatland forest type Vaccinio uliginosae-Pinetum rotundatae (see S1, T2).

Main forest types according to country schemes

Selected examples of typical associations:

ITALY: Larici-Pinetum cembrae, Vaccinio-Rhododendretum laricetosum.

SLOVENIA: Rhodothamno- Pinetum mugho, Laricetosum deciduae.

AUSTRIA: Larici-Pinetum cembrae, Pinetum cembrae, Laricetum deciduae, Lycopodio annotini-Pinetum uncinatae, Rhodothamno-Rhododendretum hirsuti, Erico carnae-Pinetum prostratae, Vaccinio myrtilli-Pinetum montanae, Erico carnae-Pinetum uncinatae, Rhododendro hirsuti-Pinetum montanae.

SWITZERLAND: Larici-Piceetum, Larici-Pinetum cembrae, (Rhododendro ferruginei-Pinetum montanae, Sphagno-Pinetum montanae Pyrolo-Pinetum montanae, Rhododendro hirsuti-Pinetum montanae).

GERMANY: Vaccinio-Pinetum cembrae, Vaccinio-Rhododendretum ferruginei.

CORINE: 3.1.2

Key factors of forest biodiversity

STRUCTURAL: Micro-site conditions (e.g. exposure, depressions) are very important to forest regeneration and hence stand structure. When compared with other forest types, tree canopies are uneven and stand density lower. Horizontal structure is generally clustered, and more significant at the upper timberline. At lower altitudes clusters are observed mainly in the regeneration phase. In later phases they fuse.

COMPOSITIONAL: At the upper timberline of the central and eastern part of the Alps, the larch and/or Swiss stone pine forests show a smooth transition to shrub vegetation of dwarf pine *Pinus mugo*. Open forests of *P. mugo* var. *uncinata* are found in the western and central part of the Alps (France, Italy, Switzerland). In the climatically more atlantic alpine and dinaric regions, coniferous forest types are replaced by beech, which form a so-called maritime forest type at the upper timberline. Further upward this is replaced by dwarf pine shrub vegetation. Larch *Larix decidua* is not present in the dinaric region, with spruce and dwarf pine forests found at much lower altitudes in deep and closed terrain depressions, where frequent thermal inversion situations ("Doline") occur.

NATURAL DISTURBANCES: This forest type is maintained by frequent natural disturbances (especially abiotic factors) such as wind throw, snow break, avalanches and soil erosion. For this reason this forest is permanently regenerating and therefore in the selection developmental phase. Disturbances of large intensity may also occur (e.g. avalanches). Subsequent progressive forest succession may therefore last for hundreds of years. Fire may also act as an important ecological factor, although this has not yet been sufficiently analysed until recently.

Current land use/silviculture (including history and trends)

Forest history, particularly settlement and agricultural land-use, has been the major factor affecting the formation of this forest type. Anthropogenic influences on this forest type have always been very strong particularly since the expansion of deforestation resulting from fire. Pasturage has been very important since earliest human presence, and has thus influenced stand structure by lowering the upper timberline by 200–400 m. Larch is particularly favoured by fire and other human interventions. According to pollen analysis, it is presumed that the current area of larch forest is much larger than it was originally. At the present time, many of these forests are managed using low intensive silvicultural practices or they are managed to protect settlements, roads and railway lines against avalanches, landslides and torrents. Grazing pressure is still, however, an important influence on regeneration and stand structure.

Subalpine conifer vegetation in nemoral zone C3, photo: Peter Friis Møller.

Subalpine conifer vegetation in nemoral zone C3

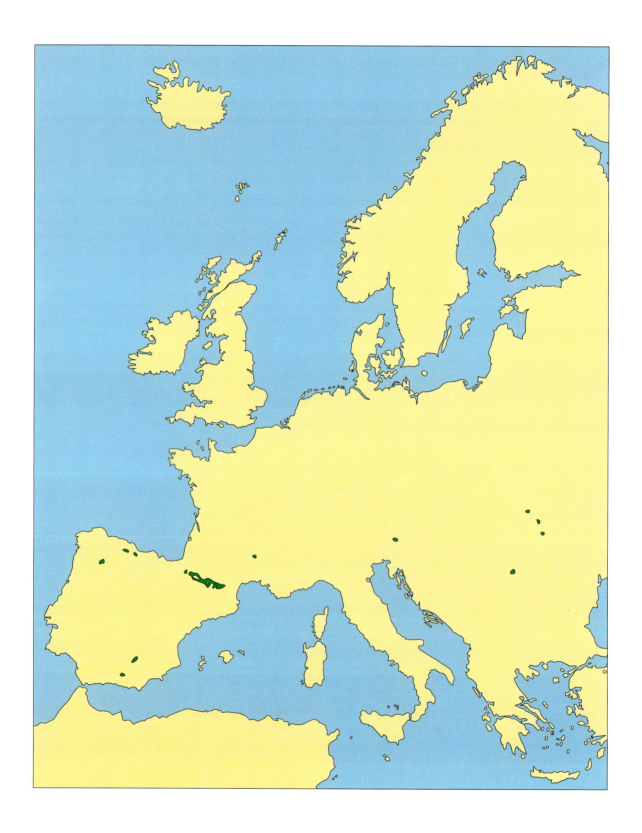

North boreal spruce forest D1

Characteristics, ecological conditions and main tree species

The north boreal forest (taiga in strict sense) is patchily distributed in northern Norway and as a wide continuous belt in NW Sweden and N Finland. Throughout Fennoscandia this forest type is dominated by *Picea abies* and with *Picea obovata* in the northernmost regions. At high altitude and latitude *Picea* reaches its distribution border and only *Betula pubescens* spp. *czerepanovii* remains. North boreal spruce forests exist on sites ranging from mesic to the most fertile and moist soils of the north boreal zone of Fennoscandia. In Finland and Norway D1 is not such an important forest type as D10 (North boreal pine forests) in the north boreal zone because of variations in site conditions. This fact is exceptional in the global scale of boreal zone, where northern regions are dominated by spruce forests. In Sweden, by contrast, D1 is more common.

Main forest types according to country schemes

The main northern spruce forest types vary among the Fennoscandian countries. While in Finland this forest type is found mostly on the poorer site types ranging from Myrtillus-type (and *Vaccinium*-type) in southern and northern Lapland (*Hylocomium*-Myrtillus-type via the *Empetrum*-Myrtillus, *Ledum*-Myrtillus to the Uliginosum-*Empetrum*-Myrtillus type (Cajander 1949, Kalela 1961), in Sweden and Norway it occurs on a much wider range of sites from *Empetrum* to tall herb (Hägglund and Lundmark 1977, Fremstad 1997).
CORINE: 3.1.2

Key factors of forest biodiversity

Under no human influence usually no large stand-replacing disturbances occur except for windfall events in the westernmost part of the region (W Norway). The main disturbance factors maintaining spruce forest biodiversity are snow-breaks, lethal frost and red rot fungi resulting in small-scaled gap-dynamics which under natural conditions usually form extensive areas of unfragmented old-growth forests. Fire, insect outbreaks or wind are not dominating disturbance factors in this forest type. However, the role of fire increases to the east and in the region formed by *Betula pubescens* spp. *tortuosa*, insect outbreaks are intensive. In this region *Oporinia* is the major disturbance factor.

STRUCTURAL: Old trees, snags and dead wood are important structures that maintain the characteristic biodiversity. The volume of dead wood is lower than in many other more productive forest types, but the proportion of dead wood of the total biomass is still high.

Forest stands can achieve a high mean age but have, due to gap-phase dynamics, wide age and diameter distributions (Kuuluvainen 1994). In the northernmost parts of Fennoscandia the natural reproduction of spruce forest can be irregular due to climatic conditions (poor seed production) and site conditions (a very thick moss layer). As a consequence many high-altitude forests have age structures in the form of damped wave with certain dominating age classes.

This means that without human interference, the parts of the landscape with moist site conditions would be dominated by continuous forest with a high mean age.

COMPOSITIONAL: This type is characterised by the dominance of *Picea obovata* in northernmost parts, but mostly dominated by *Picea abies*. As we move towards the east, Siberian fir *Abies sibirica* and larch *Larix* enter as new dominating genera (D4, E of Fennoscandia). Due to the absence of disturbances other conifer and broad-leaved tree species are mixed with *Picea*, such as *Pinus sylvestris*, *Betula pubescens* spp. *pubescens*, and *Betula pubescens* spp. *czerepanovii*, but usually suppressed due to thick moss layer (*Hylocomium splendens*, *Pleurozium schreberi*). Dwarf shrubs such as *Vaccinium myrtillus* are the most significant species in the field layer. Due to decreased evaporation towards the north, species such as *Ledum palustre* and *Vaccinium uliginosum*, which are otherwise known as mire plants, become frequent (missing in Norway).

Due to the relatively low frequency of stand replacing disturbance, old and very old stands with a specific fauna and flora are found (e.g. *Picoides tridactylus*, *Alectoria sarmentosa*, *Sceletocutis tschulymica*). In sub-alpine birch forests *Oporinia* is an important key-stone species.

Small mammals show strong inter-annual variation in numbers, which drives the dynamics of many other species, including both predators and other kinds of prey.

NATURAL DISTURBANCES: The natural disturbance with gap-phase dynamics and a low frequency of stand-scale replacement has been almost completely replaced by stand-wise harvesting by clear-cutting, followed by spruce planting.

Current land use/silviculture (including history and trends)

North boreal spruce forests have been out of intensive anthropogenic forest exploitation until the early to mid 20th century. After that and up to the 1990s forestry was quite intensive. From 50s to 80s large continuous clear-cut areas and very heavy site preparation was favoured. Recently the management has changed in favour of smaller-scale clear-cutting and locally even different forms of selective cutting.

Still, the current land use is largely dominated by the clear-cutting activities during the past 70 yr and it has strongly reduced the amount of authentic forests. Extensive reindeer grazing on tree lichens in spruce forest and ground lichens in pine forests occur.

Further subdivision of forest types

Paludified spruce forests with thin peat layer can also be classified into this forest biodiversity type.

North boreal spruce forest D1. The tall-herb vegetation can be found e.g. in slopes exposed to overflow of surface water, photo: Peter Friis Møller.

North boreal spruce forest D1 and north boreal pine forest D10

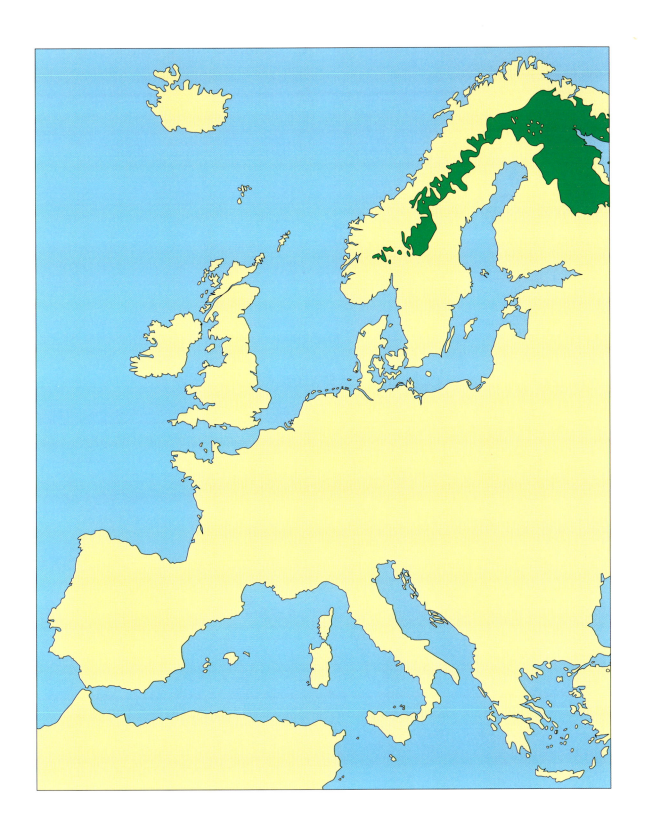

North boreal pine forest D10

Characteristics, ecological conditions and main tree species

Scots pine *Pinus sylvestris* is totally dominating in these forests. This forest type is found on nutrient poor and dry sites and its dominance increases to the east. In Finland and in Norway this type is the most dominating type of the north boreal region. On the contrary, in Sweden D1 is the most dominating (see D1).

Main forest types according to country schemes

In all Fennoscandian countries this forest type is found on the parallel site types of *Cladonia* and *Vaccinium*, (*Cladonia*-type, Uliginosum-*Vaccinium*-*Empetrum*-type, Myrtillus-*Calluna*-*Cladonia*-type) and on pine mires with thin peat layer. All these types are found in southern and northern Lapland, both in Finland and Sweden, as well as in northernmost Norway (Cajander 1949, Hägglund and Lundmark 1977, Fremstad 1997).

CORINE: 3.1.2

Key factors of forest biodiversity

Under natural conditions the key factors maintaining forest biodiversity is the interaction between the forest and fire. Fires usually have low intensities but occur at short (30 yr) intervals. As a consequence, trees from several age cohorts, large old trees as well as standing and downed dead wood are continuously present in most stands.

STRUCTURAL: Combined with the high tolerance of Scots pine to fire, the resulting stands usually contain several distinct age classes ranging from young to very old trees (Huse 1965). There is also great variation in the size of burns and in the average age distribution of stands within the landscape. Pine forest stands can achieve mean ages approaching 300 yr, which further increases the variation. Snags and dead wood are important structures. The volumes of dead wood is lower than in many other more productive forest types, but the proportion of dead wood of the total biomass is still high.

COMPOSITIONAL: Often Scots pine *Pinus sylvestris* is dominating totally in these forests. Deciduous trees like *Betula pubescens* spp. *pubescens* and in the northernmost regions of north boreal zone *Betula pubescens* spp. *czerepanovii* can form a poorly developed understorey.

North boreal pine forest D10, photo: Peter Friis Møller.

Throughout the most northern Fennoscandia birch *Betula pendula* spp. *tortuosa* is the dominating species and also forms the timber line.

The ground layer vegetation is dominated by lichens of the genus *Cladonia*, occasionally of *Stereocaulon*. Dwarf shrubs, like *Vaccinium vitis-idaea* form the field layer. *Betula nana* can also exist in the field layer.

The structurally diverse natural pine forest hosts several specialist birds (*Tetrao urogallus*, *Parus cinctus* and *Perisoreus infaustus*) as well as insects (*Tragosoma depsarium*) and fungi (e.g. *Anthrodia albobrunnea*).

NATURAL DISTURBANCES: Under natural conditions the disturbance factor maintaining forest biodiversity is the fire. Fires usually have low intensities but occur at short (30 yr) intervals. By contrast, stand-wise harvesting by clear-cutting is presently followed by scarification and natural regeneration from seed trees.

Current land use/silviculture (including history and trends)

In this forest type the large trees were selectively cut already during the 18th and 19th centuries. Today forestry is intensive on commercially exploited areas, but in northern Finland and NE Norway large areas are included within conservation areas, but with no restriction to reindeer grazing.

Further subdivision of forest types

Paludified pine forests with a thin peat layer and an abundant cover of hummock dwarf shrubs can be classified into this forest type.

North boreal spruce forest D1 and north boreal pine forest D10

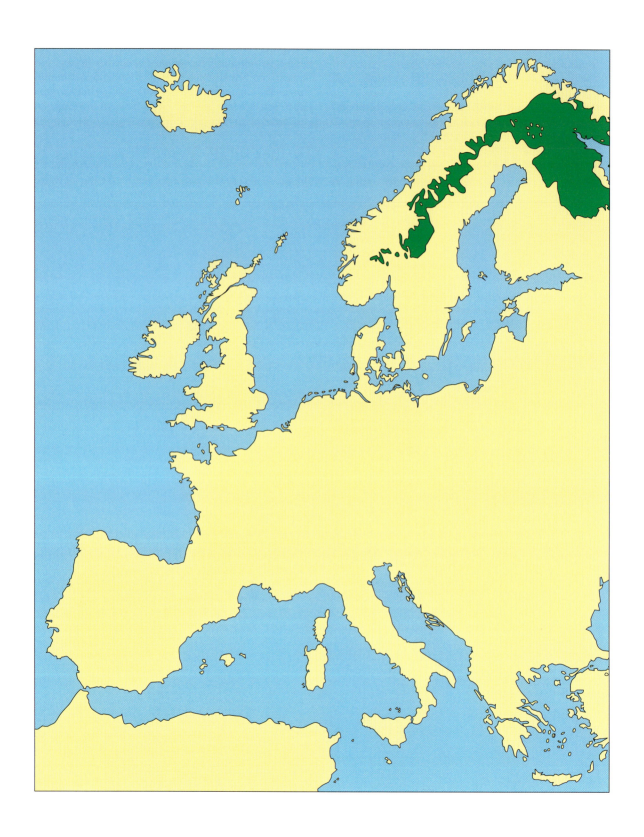

Middle boreal spruce forests D2

Characteristics, ecological conditions and main tree species

Middle boreal spruce forests are found throughout central Norway, Sweden and Finland. *Picea abies* dominates in later successional stages, with a mixture of *Pinus sylvestris* and *Betula pubescens* as remnants of the earlier succession stages. Under natural succession also other deciduous tree species may be abundant.

Main forest types according to country schemes

The main site types of spruce-dominated sites throughout the different Fennoscandian countries range from *Vaccinium*-Myrtillus to *Oxalis*-Myrtillus types typical for the middle boreal region (Cajander 1949, Kalela 1961, Hägglund and Lundmark 1977, Fremstad 1997).
CORINE: 3.1.2

Key factors of forest biodiversity

Under natural conditions the key factors maintaining forest biodiversity is the interaction between the forest and fire as well as to some extent the effect of strong wind. Fires usually have low intensities but occur at short (30 yr) intervals. There is usually a continuous presence of large old trees as well as standing and downed dead wood.

STRUCTURAL: At the stand scale naturally dynamic forests show large differences between different successional stages after stand replacing disturbance. During early- and mid-successional stages (approaching 100 yr) deciduous trees dominate but are then disappearing due to a relatively short life span. In late successional stages *Picea abies* takes over and may develop into multi-storey old-growth stands with small-scaled gap dynamics. The amount and types of dead wood also changes with succession. After stand-replacing disturbances the amount of dead wood is close to the volume found in the previous stand and burned wood common after fire. During early succession the remnant dead wood decays over the first decades. Then there is a gradual build-up of dead wood both in the canopy as snags and on the ground as woody debris. In very old stands the amount of dead wood may amount to 20–40% of the total wood biomass. On the most fertile sites the amount of coarse woody debris can be remarkable and exceed 100 m³ ha⁻¹. At the landscape scale the natural stand sizes show large variation ranging from <1 to >10 000 ha. Typically the age distribution has a wide range of age classes from recent burns and deciduous successions to old-growth forests with a stand age well over 200 yr.

COMPOSITIONAL: This type is found on mesic to fertile and moist site types of the middle boreal zone of Fennoscandia. It is characterised by an increasing dominance of *Picea abies* with age with an admixture of deciduous trees, such as *Betula* spp., *Populus tremula*, *Alnus* spp. and *Sorbus aucuparia* in early to mid-successional stages. The variation in fertility of the soil can be seen in the composition of ground layer vegetation. The moss layer (*Pleurozium schreberi*, *Rhytidiadelphus triquetrus*) may not be continuous as in more northern types, and lichens are almost totally missing. On the most fertile sites species like *Oxalis acetosella* and *Maianthemum bifolium* are typical in the field layer. As a rule the field layer is dominated by *Vaccinium myrtillus*, but in younger successional stages it is more characterised by grasses (*Calamagrostis* spp., *Deschampsia flexuosa*) and herbs. On less fertile sites the dominance of *V. myrtillus* increases and the abundance of grasses and herbs decreases. The shrub layer is often formed by *Sorbus aucuparia* and *Salix* spp.

Different species have adapted to living in the succession from a stand-replacing disturbance by fire to old-growth. The following stages and associated species can be identified (e.g. Angelstam 1998a, b).
1. Recently burned ground and wood: vascular plants (*Geranium bohemicum/lanuginosum*) and pyrophilous insect species.
2. Young forest with a well-developed field layer of ericaceous species (used by *Tetrao tetrix*, *Alces alces*).
3. Deciduous succession with increasing amounts of dead wood with age: lichens (*Lobarion*), insects and birds specialising on the deciduous component (*Bonasa bonasia*, *Aegithalos caudatus*) and in combination with dead wood (*Dendrocopos leucotos* and *D. minor*).
4. Old/ageing forests with pine: birds (*Tetrao urogallus*).
5. Old-growth forest very old trees and with large amounts of dead wood: birds (*Picoides tridactylus*); lichens (*Usnea* spp., *Alectoria* spp.).

NATURAL DISTURBANCES: The major natural disturbance factor is fire. The relative importance of different natural disturbances show a clear west-east trend. In the western part of each country, forests are more vulnerable to wind disturbances and affected by the Atlantic climate. By contrast, fire becomes relatively more important towards the east of each country. The intensity of insect outbreaks and fungal diseases is moderate throughout these region. The most fertile/wet sites can form no-fire refugia, and develop wide variation in diameter and age distribution due to self-thinning and small-scaled gap-dynamics resembling the north boreal spruce forest. Snags and dead wood are important components throughout the succession. During the last decades the natural dynamics has been replaced by intensive forest management, which has decreased the within-stand variation with relation to structure and composition as well as decreased the variation of age classes at the landscape scale.

Current land use/silviculture (including history and trends)

Forestry has locally been intensive in regions with charcoal and tar production during several centuries, and has been intensive since the early to mid-19th century when international timber demand reached Fennoscandia. Along the coasts this forest type has been fragmented by clearing for

agricultural use since the past 2000 yr or more. Due to a long history of agricultural developement this forest type has disappeared from the most fertile and wet sites and is found mainly on mesic sites in Sweden and Finland. By contrast, a gap-phase dominated variant of this forest type is found in west-central Norway. After WW2 forests have been affected extensively by clear-cutting which has gradually changed the age distribution in the landscape and has reduced the amount of natural forest remnants. During the 1990s there has, however, been a shift towards increased nature considerations.

Further subdivision of forest types

The paludified spruce forests with thin peat layer can be classified into this forest type. This dominating forest type of the region (successional stages of D2) shifts towards east of the Fennoscandian shield to D5.

Middle boreal spruce forests D2, photo: Peter Friis Møller.

Middle boreal spruce forest D2 and middle and south boreal and hemiboreal pine forest D11

Middle and south boreal and hemiboreal pine forest D11

Characteristics, ecological conditions and main tree species

This forest type is found on nutrient poor and/or dry sites of the region. These sites may be the ones with a shallow soil layer on the crystalline bedrock as well as on deep sandy and gravely soils. The field layer types range from lichen and lichen-rich to those dominated by *Vaccinium vitis-idea* and *Arctostaphylos uva-ursi*.

Pinus sylvestris is the dominating tree species in these forests, but *Betula verrucosa*, *Betula pubescens*, and *Sorbus aucuparia* may exist as an admixture on more fertile sites. In the hemiboreal zone the number of southern plant species, which are totally missing in south and middle boreal zones, increases.

Main forest types according to country schemes

Throughout Fennoscandia this forest type is found on site types ranging from pure lichen, *Empetrum*, *Calluna* and *Vaccinium*. On the driest, rocky sites with shallow soils, lichens are the dominating ground vegetation. Dwarf shrubs (e.g. *Calluna vulgaris*, *V. vitis-idaea*) are common but do not form a dense vegetation community. On these sites hardwood tree species are often poorly represented. On more fertile types the number and density of dwarf shrubs increases and hardwood tree species can reach the tree layer. A typical variant is the pine forests on shallow peaty soils with a predominance of *Ledum palustre* in the field layer.

CORINE: 3.1.2

Middle and south boreal and hemiboreal pine forest D11, photo: Tor-Björn Larsson.

Key factors of forest biodiversity

Under natural conditions the key factors maintaining forest biodiversity is the interaction between the forest and fire. Fires usually have low intensities but occur at short (30 yr) intervals. There is usually a continuous presence of large old trees as well as standing and downed dead wood.

STRUCTURAL: Combined with the high tolerance of Scots pine to fire, the resulting stands usually contain several distinct age classes ranging from young to very old trees. There is also great variation in the size of burns and in the age distribution within the landscape. Pine forest stands can achieve mean ages approaching 300 yr, which further increases the variation. Snags and dead wood are important structures. The volumes of dead wood is lower than in many other more productive forest types but the proportion of dead wood of the total biomass is still high.

COMPOSITIONAL: Often *Pinus sylvestris* is totally dominating in these forests. The ground layer vegetation is dominated by lichens of the genus *Cladonia*, occasionally of *Stereocaulon*. Dwarf shrubs, like *Vaccinium vitis-idaea* form the field layer.

The structurally diverse natural pine forest host several specialist birds (*Tetrao urogallus*, *Caprimulgus europaeus*, *Lullula arborea*) as well as insects (*Tragosoma depsarium*) and fungi (*Albobrunnea lutea*).

NATURAL DISTURBANCES: The key factors maintaining forest biodiversity is the interaction between the forest and fire. Fires usually have low intensities but occur at short (30 yr) intervals. The pines tolerate fire well and the density of pine forests is often low, which leads to ground fires and to great variation in size and age distribution within the forest. This means that the successional stages have surviving remnant trees from the previous stand. By contrast, stand-wise harvesting by clear-cutting is presently followed by scarification and natural regeneration from seed trees.

Current land use/silviculture (including history and trends)

In this forest type the large trees were selectively cut already a very long time ago. Consequently natural remnants are very rare. Today forestry is intensive on commercially exploited areas.

Middle and south boreal and hemiboreal burned pine forest D11, photo: Tor-Björn Larsson.

Middle boreal spruce forest D2 and middle and south boreal and hemiboreal pine forest D11

South boreal forest D3

Characteristics, ecological conditions and main tree species

This type is found in south-central Norway, Sweden and Finland. It is characterised by more or less the same features as found in the middle boreal spruce zone, i.e. by an increasing dominance of *Picea abies* with age, and with an admixture of deciduous trees, such as *Betula* spp., *Populus tremula*, *Alnus* spp. and *Sorbus aucuparia* in early- to mid-successional stages.

Main forest types according to country schemes

The main site types throughout the different Fennoscandian countries range from *Vaccinium*-Myrtillus to *Oxalis*-Myrtillus types typical for the south boreal region (Cajander 1949, Kalela 1961, Hägglund and Lundmark 1977, Fremstad 1997).

CORINE: 3.1.2

Key factors of forest biodiversity

Under natural conditions the key factors maintaining forest biodiversity is the interaction between the forest and fire as well as to some extent the effect of strong wind, especially in western Fennoscandia. The detailed fire history is more difficult to assess than in the more northern forest types where fire-scarred trees are still found and can be analysed. The occurrence of distinct successional stages is an important feature. During natural succession there is usually a continuous presence of large old trees as well as standing and downed dead wood.

STRUCTURAL: At the stand scale, naturally dynamic forests show large differences between different successional stages after stand replacing disturbance. During early- and mid-successional stages (approaching 100 yr) deciduous trees dominate but are then disappearing due to a relatively short life span. In late successional stages *Picea abies* takes over and may develop into multi-storey old-growth stands with small-scaled gap dynamics. The amount and types of dead wood also changes with succession. After stand-replacing disturbances the amount of dead wood is close to the volume found in the previous stand and burned wood common after fire. During early succession the remnant dead wood decays over the first decades. Then there is a gradual build-up of dead wood both in the canopy as snags and on the ground as woody debris. In very old stands the amount of dead wood may amount to 20–40% of the total wood biomass.

At the landscape scale the natural stand sizes show large variation ranging from <1 to >10 000 ha. Typically the age distribution has a wide range of age classes from recent burns and deciduous successions to old-growth forests with a stand age well over 200 yr.

COMPOSITIONAL: The fertility can vary quite a lot, being partly in parallel with the middle boreal zone types. The field layer is characterised by more diverse vegetation of grasses and herbs (e.g. *Anemone nemorosa*, *Pyrola* spp., *Hepatica nobilis*, *Melica nutans*, *Solidago virgaurea*) than the more northern types. The shrub layer is more developed compared with the middle boreal zone with *Ribes* spp., *Rhamnus frangula* and *Rubus idaeus*. This type is characterised by an increasing dominance of *Picea abies* with age with an admixture of deciduous trees, such as *Betula* spp., *Populus tremula*, *Alnus* spp. and *Sorbus aucuparia* in early- to mid-successional stages. On the most fertile sites scattered isolated remnants from a previously wider distribution range of temperate broad-leaved tree species, such as lime *Tilia cordata*, maple *Acer platanoides*, elm *Ulmus glabra* and ash *Fraxinus excelsior* can be found.

Different species have adapted to living in the succession from a stand-replacing disturbance by fire to old-growth. The following stages and associated species can be identified (e.g. Angelstam 1998a, b).

1. Recently burned ground and wood: vascular plants (*Geranium bohemicum/lanuginosum*) and pyrophilous insect species.
2. Young forest with a well-developed field layer of ericaceous species (used by *Tetrao tetrix*, *Alces alces*).
3. Deciduous succession with increasing amounts of dead wood with age: lichens (*Lobarion*), insects and birds specialising on the deciduous component (*Bonasa bonasia*, *Aegithalos caudatus*) and in combination with dead wood (*Dendrocopos leucotos* and *D. minor*).
4. Old/ageing forests with pine: birds (*Tetrao urogallus*).
5. Old-growth forest with very old trees and with large amounts of dead wood: birds (*Picoides tridactylus*); lichens (*Usnea* spp., *Alectoria* spp.).

NATURAL DISTURBANCES: The relative importance of different natural disturbances show a clear west-east trend. In the western part of each country, forests are more vulnerable to wind disturbances and affected by the Atlantic climate. By contrast, fire becomes relatively more important towards the east of each country. The intensity of insect outbreaks and fungal diseases is moderate throughout these region.

The most fertile/wet sites probably formed no-fire refugias with wide variation in diameter and age distribution due to self-thinning and small-scaled gap-dynamics resembling the north and middle boreal spruce forest. However, this forest type has been fragmented by clearing for agricultural use since the past 2000 yr or more. Due to a very long history of agricultural development this forest type has disappeared from the most fertile and wet sites and is now found mainly on mesic sites in Norway, Sweden and Finland. As a consequence, the site types available for small-scaled gap-dynamics has been seriously reduced.

Current land use/silviculture (including history and trends)

In eastern Finland slash-and-burning cultivation was applied up to 1920s (Heikinheimo 1915). This has led to very fertile sites dominated by *Betula verrucosa*. However,

land use has changed and spruce is replacing degenerating birch coverage. Forestry has locally been intensive in regions with charcoal and tar production during several centuries, and has been intensive since the early to mid-19th century when international timber demand reached Fennoscandia. After WW2 forests have been affected extensively by clear-cutting which has gradually changed the age distribution in the landscape and has reduced the amount of natural forest remnants. During the 1990s there has, however, been a shift towards increased nature considerations.

Further subdivision of forest types

The paludified spruce forests with thin peat layer can be classified into this forest type. This dominating forest type of the region (successional stages of D3) shifts towards east of the Fennoscandian shield to D6. Locally, northern outposts of the hemiboreal forest elements can be found.

South boreal forest D3, photo: Peter Friis Møller.

South boreal forest D3

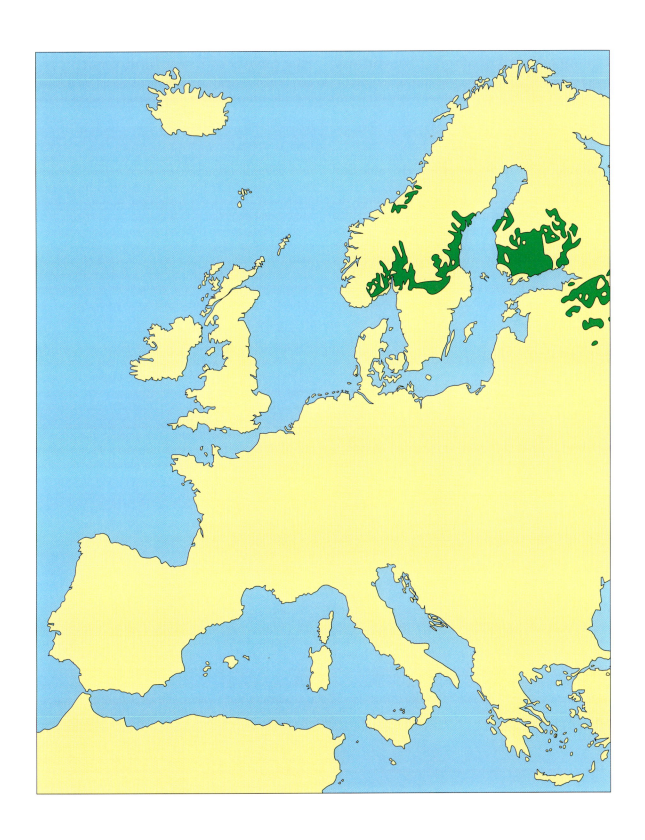

Hemiboreal spruce and fir-spruce forest D8+D12

Characteristics, ecological conditions and main tree species

The hemiboreal forests form a complex mixture of boreal and nemoral forest types and of the old cultural landscape. Along with Scots pine and Norway spruce, there are several deciduous species including *Betula* spp., *Populus tremula*, *Salix* spp. and broad-leaved trees such as *Quercus*, *Fraxinus* and *Tilia*.

In Fennoscandia it is found mainly in south-central Sweden, south of the distinct biogeographical transition zone Limes Norrlandicus which constitutes the southern border of the Eurasian taiga (Fransson 1965). In Norway and Finland this forest region is represented only as narrow bands in southernmost parts of the countries. East of the Baltic Sea it is the dominating forest type covering most of Estonia, Latvia, Lithuania, NE Poland, Belarus and Russia.

Main forest types according to country schemes

There was until recently no established typology for hemiboreal forests. Forests types have thus to be described according to management and exploitation level. It is possible to discern original hemiboreal structures, mixed forests after abandonment of wooded meadows, and dominating conifer monocultures. In Sweden the new classification presented in the 1970s also covers hemiboreal forests (Hägglund and Lundmark 1977). Original hemiboreal forest with large conifers and southern deciduous trees occur almost only in reserves such as the representative and famous Białowieża National Park in eastern Poland. Other reserves are more limited in size. Productive areas of hemiboreal forests were early used in the infields of agricultural areas as wooded meadows. Ash, oak and hazel were typical trees and bushes. There are now many abandoned sites in Sweden, Finland and the Baltic States with conifer, particularly spruce, invasion. Outfields with mixed conifers and deciduous trees (including earlier regularly burned heathland) have been transformed to conifer monocultures, with pine or pine-spruce forest on sand or bedrock and spruce on more fertile soil. Such plantations have already been regenerated once or twice.

CORINE: 3.1.2

Key factors of forest biodiversity

Due to the long land use history it is doubtful if it is meaningful to relate the changes in relation to natural conditions only. Instead, the vast majority of the landscape was a cultural landscape where forests were strongly affected by grazing, pollarding, lopping and other kinds of local forest use. In some regions (southwestern Sweden) the forest cover disappeared completely and was replaced by extensive *Calluna* heaths. The key factors maintaining forest biodiversity is therefore a complicated mixture of natural and cultural disturbances. These range from strong effects of wind in the west and the interaction between the forest and fire on drier sites in the east. However, all these disturbances did maintain a continuous presence of large old trees as well as standing and downed dead wood even though old-growth forests in the natural sense were not common (see also the forest type open woodland).

STRUCTURAL: At the stand scale the dynamics of these forests produced large differences between different successional stages after stand-replacing disturbance and the traditional land use. During early- and mid- successional stages (approaching 100 yr) deciduous trees dominate but are then disappearing due to a relatively short life span. In late successional stages *Picea abies* takes over and may develop into multi-storey old-growth stands with small-scaled gap dynamics. The amount and types of dead wood also changes with succession. After stand-replacing disturbances the amount of dead wood is close to the volume found in the previous stand and burned wood common after fire. During early succession the remnant dead wood decays over the first decades. Then there is a gradual build-up of dead wood both in the canopy as snags and on the ground as woody debris. In very old stands the amount of dead wood may amount to 20–40% of the total wood biomass.

The cultural landscape also maintained a continuous presence of large old trees as well as standing and downed dead wood even though old-growth forests in the natural sense were not common (see also the forest type open woodland).

COMPOSITIONAL: This type is found in mesic to fertile and moist site types in southern Sweden and southernmost Norway and Finland. It is characterised by an increasing dominance of *Picea abies* with age with an admixture of deciduous trees, such as *Betula* spp., *Populus tremula*, *Alnus* spp. and *Sorbus aucuparia* in early- to mid-successional stages. On the most fertile sites scattered isolated remnants from a previously wider distribution range of temperate broad-leaved tree species, such as lime *Tilia cordata*, maple *Acer platanoides*, elm *Ulmus glabra*, ash

Fraxinus excelsior and oak *Quercus robur* can be found. Different species which have adapted to living in the succession from a stand-replacing disturbance by fire to old-growth can be found, partly because of an intact natural dynamics, and partly because of the long presence of a cultural landscape. The following stages and associated species can be identified (e.g. Angelstam 1998a, b).

1. Recently burned ground and wood: vascular plants (*Geranium bohemicum/lanuginosum*) and pyrophilous insect species.
2. Young forest with a well-developed field layer of ericaceous species (used by e.g. *Tetrao tetrix*, *Alces alces*)
3. Deciduous succession with increasing amounts of dead wood with age: lichens (*Lobarion*), insects and birds specialising on the deciduous component (*Bonasa bonasia*, *Aegithalos caudatus*) and in combination with dead wood (*Dendrocopos leucotos* and *D. minor*).
4. Old/ageing forests with pine: birds (*Tetrao urogallus*).
5. Old-growth forest very old trees and with large amounts of dead wood: birds (*Picoides tridactylus*); lichens (*Usnea* spp.).

Natural disturbances: Cultural management, wind (in the west) and fire (in the east) can be considered as the major disturbance factors.

Current land use/silviculture (including history and trends)

In Norway and Sweden this forest type has been used for thousands of years, cf. above.

Further subdivision of forest types

One exceptional feature of this forest biodiversity type is the prairie-alike vegetation of southern Sweden in regions with alkaline base rock. These areas are almost open, including a poorly developed pine coverage with *Juniperus communis*. Famous type of vegetation is *Helianthemum oelandicum* heaths ("Alvar") of the Öland-island, which is endemic to this region.

Hemiboreal spruce and fir-spruce forest D8+D12, photo: Peter Friis Møller.

Hemiboreal spruce and fir-spruce forest D8+D12

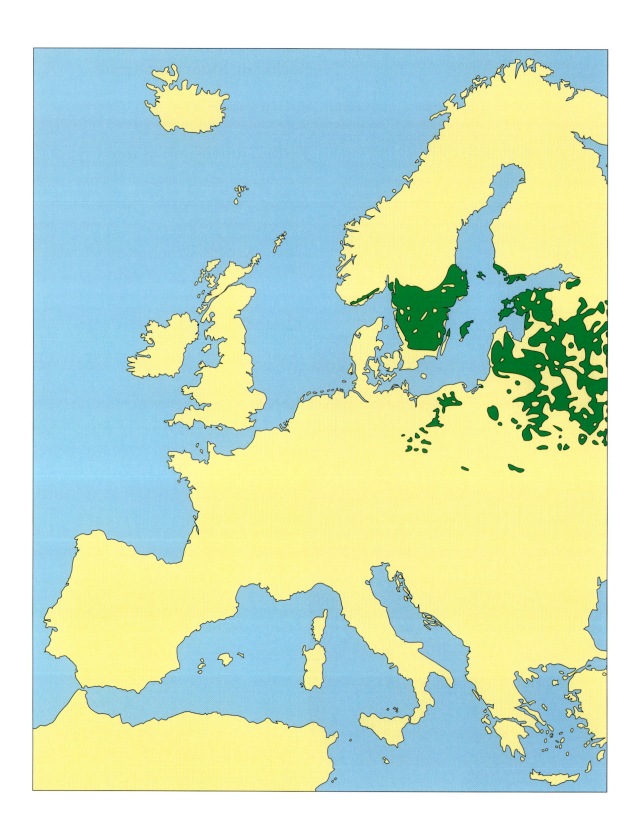

Mixed spruce and fir forest D9

Characteristics, ecological conditions and main tree species

This natural forest type is present both in the Alpine region and in mountainous regions of central Europe (Bayerischer Wald, Harz, Vogesen, Schwarzwald). Similar forests cover areas of the Carpathians. However, the latter sub-types are not described in this text, which focuses on the alpine and central European types. Many potential mixed forest types are presently being transformed into pure spruce-forests, which are similar to the natural spruce forests. These forest types will, however, be treated separately from a biodiversity viewpoint.

In the NORTHERN ALPS, a clear gradient exists from the northern most peripheral Alps to the central Alps in the mountainous altitudinal zone (ca 600–1600/1850 m a.s.l.), as a result of climatic conditions:

- Peripheral Alps: subatlantic influence of climate; bedrock mainly limestone; mixed beech-fir-spruce forests.
- Transition zone of the Alps: transitional climate (between subatlantic and continental); bedrock mainly silicate; mixed fir-spruce forests. Almost all of these potentially well balanced spruce-fir natural forests have been transformed by intensive forest management and deerbrowsing to nearly pure spruce forests.
- Central Alps: continental climate; silicate bedrock; pure spruce forests.

In the SOUTHERN ALPS, similar to the more complex distribution of beech-dominated forests, a less clear zonation exists.

For both geographical parts of the Alps, a spruce-dominated subalpine altitudinal belt exists above the mountainous zone. Solely in the central Alps, a high-subalpine belt of mixed stone-pine-larch forests (*Pinus cembra-Larix decidua*) exists above this spruce dominated belt (see major forest type C3). In general, from west to east, a clear gradient in the proportions of silver fir to spruce forest exists in the Alps, due to forest history and different species migration patterns. Spruce dominates the eastern Alps, and silver fir the west. Therefore, in the south-western Alps near to the Mediterranean Sea pure fir-forests occur, even close to the upper timberline (Mayer 1984).

Main forest types according to country schemes

Distribution of natural types of spruce and silver fir-spruce forests: Switzerland, Italy, France, Austria, Germany, Liechtenstein, Slovenia, Czech Rep., Slovakia, (similar types in Poland, Ukraine, Bulgaria, Romania, Greece).

Due to the favourable timber characteristics of spruce, this species has been planted on many sites for which potential dominants are silver fir or broad-leaved species.

Selected typical examples of associations:

ITALY: Picetum subalpinum, Veronica u.-picetum (Picetum montanum).

SLOVENIA: Galio rotundifolii-Abietetum, Adenostylo glabrae-Piceetum, Luzulo nemorosae-Piceetum, Avenello flexuosae-Piceetum.

AUSTRIA: Larici-Piceetum, Calamagrostio variae-Piceetum, Adenostylo glabrae-Piceetum, Luzulo nemorosae-Piceetum, Soldanello montanae-Piceetum, Veronico latifoliae-Piceetum, Adenostylo glabrae-Abietetum, Galio rotundifolii-Abietetum, Adenostylo alliariae-Abietetum, Equiseto sylvatici-Abietetum.

SWITZERLAND: Bazzanio-Abietetum, Calamagrostio villosae-Abietetum, Dryopteridio-Abietetum, Equiseto-Abietetum, Adenostylo-Abietetum, Galio-Abietetum.

GERMANY: Luzulo-Abietetum, Vaccinio-Abietetum, Galio rotundifoliae-Abietetum, Pyrolo-Abietetum, Bazzanio-Piceetum (see S1). Vegetation character of ground vegetation is acidophytic in Vaccionio-Abietetum and Luzulo-Abietetum but mesophytic to calcarophytic in Galio-Abietetum and Pyrolo-Abietetum; that is why in the phytosociological classfication the latter types are incorporated in the beech forest types (F5b).

FRANCE: Luzulo-Abietetum, Oxali-Abietetum, Galio rotundifolii-Abietetum, Trochiscantho-Abietetum, Adenostylo-Abeitetum Veronico urticifoliae-Piceetum, Homogyno-Piceetum.

CORINE: 3.1.2

Mixed spruce and fir forest D9, photo: Peter Friis Møller.

Key factors of forest biodiversity

STRUCTURAL: A clear distinction has to be made between the montane and subalpine forest types, which are structurally different. Subalpine types, close to the upper timberline have a more clustered structure. The horizontal structure is clustered due to permanent regeneration, whilst vertical structure is more open and differentiated. Montane forest types are in most cases closed, with high stem density but a more homogeneous horizontal structure. Further differences in stand structure are caused by the proportions of shade tolerant silver fir, which occurs partly in montane, but not subalpine sub-types.

COMPOSITIONAL: This forest type is dominated by spruce. In general the proportion of spruce has increased mainly as a result of silvicultural treatment (e.g. clearcutting), and as a result of the decreased competitiveness of fir resulting from browsing by roe and red deer. In many regions, the silver-fir dominated subtype has nearly disappeared and species diversity and composition declined into pure spruce stands. Broad-leaved species rarely occur in this forest type. Browsing, pasturing and clearcutting have increased the competetive ability of larch *Larix decidua*, which now has extended beyond its potential natural contribution in the stand.

NATURAL DISTURBANCES: There is a clear difference between montane and subalpine types. Towards the upper timberline regeneration becomes more and more difficult. Only the most favourable micro-sites (some square metres or less) can support seedlings. Dead wood is very important to the natural regeneration of this forest type, as seeds and seedlings are likely to find adequate growth conditions on decaying wood. Because of permanent small-scale disturbance and regeneration in the sub-alpine zone, the pattern of development phases is on a much finer scale than in mountainous forest types. These are instead often affected by large windfalls and snowbreakages due to their even-aged structure, which has derived from the silvicultural treatments adopted during the last decades.

Current land use/silviculture (including history and trends)

Close to the upper timberline, and mainly in the sub-alpine forest types, pastural, browsing and clearcutting activities have allowed the successful introduction of larch. Due to silvicultural treatment and high deer populations, fir-dominated montane stands have been transformed into pure spruce stands in the transition zone (which was natu-rally fir-dominated). In many regions silver fir has nearly disappeared. Large proportions of the subalpine subtype, particularly at steep terrain or in harsh climatatic conditions, are managed mainly against avalanches, rockslides or torrents.

In comparison, most of the montane sub-types are subject to intensive forest management and used as highly productive forests. Natural mixed spruce-fir forest types originally have been dominated by silver fir but recently the relative proportion of spruce may be largely dominant. Natural spruce forests are, however, restricted to peatland and steep rocky site types (Bazzanio-Piceetum; see S1). Anthropogenic spruce forests (see spruce plantations) may today occur on almost all site types. A new silvicultural trend is to choose suitable site types for intensively managed spruce stands and to add at least 30% broad-leaved tree species, usually beech if a beech forest type is the natural potential vegetation. Other coniferous components are silver fir and increasingly Douglas fir.

Mixed spruce and fir forest D9

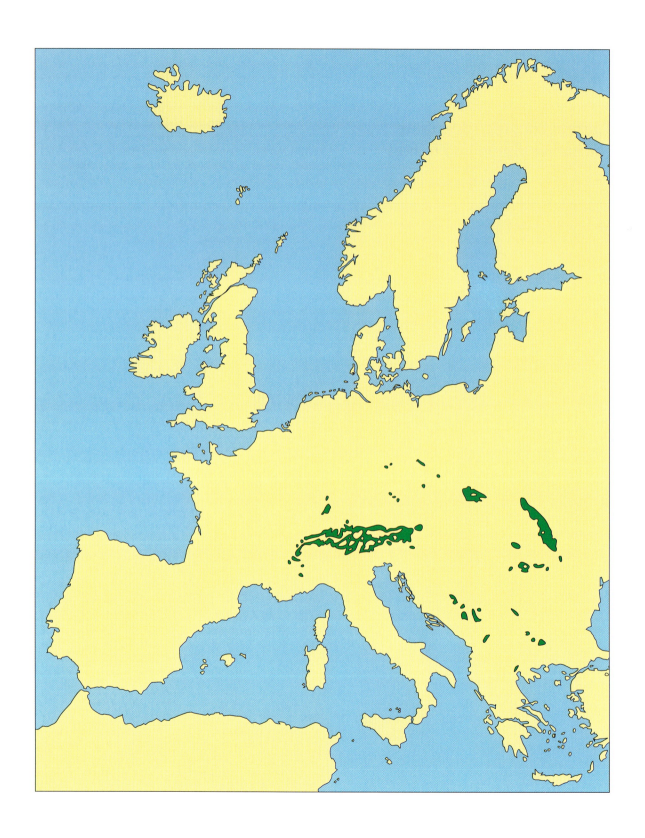

Mixed oak forest F1

Characteristics, ecological conditions and main tree species

Oak-birch forests grow on poor acidic soils (sands or rock) in Atlantic climates and colline to submontane levels. They are frequent in the northern German lowlands (Pleistocene deposits) and in the southern regions on silicate rocks, and have been managed for coppice, grazing and other human needs for a long time. Degradation of the upper soil layer and declines in forest yield were provoked due to poor nutrient supply. Generally, a poor herb layer exists and production levels are low. In the western part of Britain and Ireland, often on exposed locations and at higher elevations. In Ireland there are two types of mixed oak forest within this category namely, species poor *Quercus petraea* found principally in the drier eastern parts where rainfall is < 1200 mm, *Q. petraea* forest rich in bryophytes and occurs in the wetter parts of the country where rainfall exceeds 1200 mm (Cross 1998). In the boreal area on fertile lowland areas of the hemiboreal region. In Finland only in a few areas in the southwesternmost part of the country. In the perialpine regions

Mixed oak forest F1, photo: Peter Friis Møller.

and low-level alpine valleys, this forest type has almost disappeared. Rich in lichens and bryophytes.

MAIN TREE SPECIES: *Quercus petraea, Q. robur, Betula pendula, B. pubescens, Sorbus aucuparia, Crataegus monogyna, Fraxinus excelsior, Vaccinium* spp*., Frangula alnus, Populus tremula, Betula carpatica*. In the boreal area dominated by the temperate broad-leaved tree species, such as lime *Tilia cordata*, maple *Acer platanoides*, elm *Ulmus glabra*, ash *Fraxinus excelsior*. In Ireland *Ilex aquifolium* is dominant in the shrub layer.

Main forest types according to country schemes

Selected examples of typical associations:

GERMANY: Luzulo-Quercetum (= Quercetum medioeuropaeum) Holco mollis-Quercetum (robori-petraeae) including Betulo-Quercetum petraeae, Genisto tinctoriae-Quercetum.

HUNGARY: Luzulo-Quercetum.

IRELAND: Species-poor *Quercus petraea* forests: Blechno-Quercetum typicum and *Quercus petraea* forest rich in bryophytes and liches Blechno-Quercum scapanietosum.

AUSTRIA: Deschamsio flexuosae-Quercetum, Cytiso nigricantis-Quercetum, Molinio arundinaceae-Quercetum.

SWITZERLAND: Quercetea roburi-petraeae.

SLOVENIA: Melampyro vulgati-Quercetum petraeae.

ITALY: Probably no occurrence, see Atlantic countries.

U.K.: *Quercus-Betula-Dicranum* (W17, Rodwell 1991), *Quercus-Betula-Oxalis* (W11, Rodwell 1991) *Fraxinus-Sorbus* (W9, Rodwell 1991).

DENMARK: *Quercus-Corylus* dominant, *Quercus-Populus tremula*.

BELGIUM: Betulo-Quercetum roboris, Violo-Quercetum roboris, Fago-Quercetum petraeae, Luzulo-Quercetum. Querco petraeae Betulateum, Trientalo-Quercetum roboris and Sileno-Quercetum petraeae.

FRANCE: Oak and mixed oak forests, poor in species: Quercion robori-petraeae.

NETHERLANDS: Acidophilus oak/mixed oak: Dicrano-Pinion and Quercion robori-petraeae p.p.

CORINE: 3.1.1

Key factors of forest biodiversity

NATURAL DISTURBANCES: In central Europe most stands are subject to succession processes which ultimately culminates in a beech forest climax. Original (natural) stands are currently only present close to wet and poor peatland sites (birch carrs) and rocks. These transitional forests are characterized by the presence of *Betula pubescens* and *Betula carpathica*. Heavy grazing, leading to lack of woody understorey and a groundflora dominated by bryophytes.

Current land use/silviculture (including history and trends)

Formerly managed as coppice woods. Neglect, plus overgrazing, leading to lack of regeneration. Many now protected as nature reserves. Has nearly disappeared in the perialpine and alpine regions because of change of land use. In Ireland, in addition to grazing and lack of management and *Rhododendron ponticum* invasion is an important issue in the richer forests.

Mixed oak forest F1

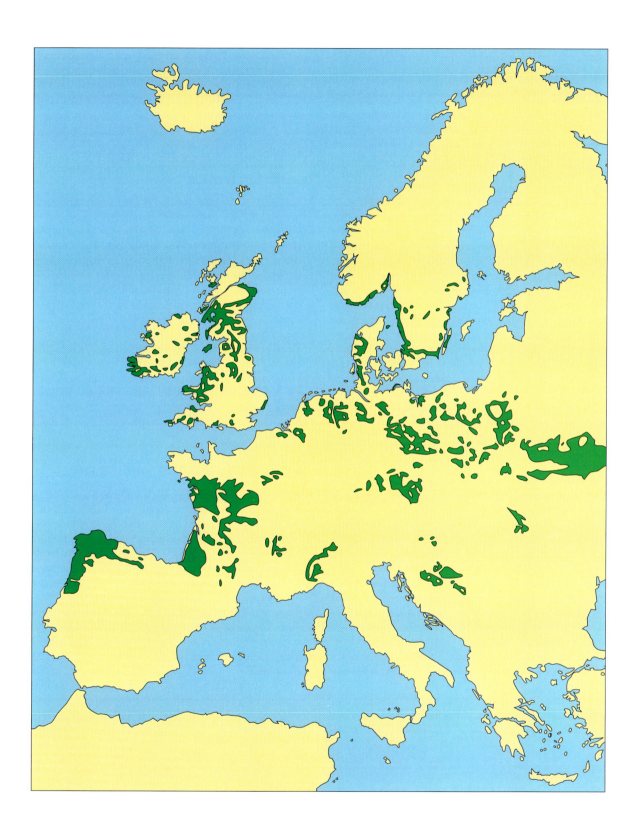

Ashwood F2

Characteristics, ecological conditions and main tree species

Atlantic ash woodlands can be divided into upland and lowland types, broadly corresponding to a northern and western distribution and a southern and eastern distribution. They grow mostly on neutral and alkaline, often moist soils, in a cool, wet and windy climate. They are particularly characteristic of limestone districts, where they form the most abundant type of semi-natural woodland. Wherever they occur, they form the richest assemblages of flowering plants and often include rare and colourful species. They also harbour a rich invertebrate fauna, and the alkaline bark of ash supports an important lichen flora.

Upland ashwoods may contain a great diversity of tree species towards the southern end of their range, in Wales and SW England. This often includes small-leaved lime *Tilia cordata*, field maple *Acer campestre*, whitebeam *Sorbus aria*, yew *Taxus baccata* and holly *Ilex aquifolium*. At the northern and oceanic end of their range, the ash component is accompanied by rowan *Sorbus aucuparia* and birch *Betula* spp. Some woods throughout their range have been invaded by sycamore *Acer pseudoplatanus* and beech *Fagus sylvatica*.

Lowland semi-natural ashwoods are concentrated on base-rich soils: rendzinas and brown earth soils developed over chalk and limestones, and other alkaline or mildly acid clays and loams. They may be sub-divided into ash-beech woods and ash-oak woods, the latter occupying the more "mesic" soils (i.e. neither the extremely dry sites on limestone outcrops, nor the acid, podzolic soils associated with heathland).

Ash-beech woods typically occur on freely-drained calcareous soils on sloping sites. Beech is characteristically accompanied by ash, wild cherry *Prunus avium*, field maple, wych elm *Ulmus glabra*, pedunculate oak *Quercus robur*, sessile oak *Quercus petraea*, and locally small-leaved lime and large-leaved lime *Tilia platyphyllos*. Where beech is not absolutely dominant and the soil is shallow, a diverse understorey of yew, holly, whitebeam and other shrubs develops. On the deeper soils, hazel *Corylus avellana* and hawthorn *Crataegus monogyna* appear. Sycamore is well-established in many woods. Dog's mercury *Mercurialis perennis* commonly carpets the ground.

Main tree species: *Fraxinus excelsior, Acer campestre/Sorbus aucuparia, Tilia cordata, (Alnus glutinosa), (Fagus sylvatica), Quercus robur, Q. petraea, Ulmus glabra, Sorbus terminalis.*

In Ireland, depending on particular sub-type (Cross 1998): dominant tree species: *Fraxinus excelsior, Quercus petraea, Quercus robur, Corylus avellana, Betula pubescens,* *Sorbus aucuparia*. Dominant shrub species: *Corylus avellana, Sorbus aucuparia, Crataegus monogyna, Ilex aquifolium*. Dominant herb species: *Hyacinthoides non-scripta, Pteridium, Hedera helix, Sesleria caerulea, Asplenium trichomanes,* and various bryophytes.

Main forest types according to country schemes

U.K.: *Fraxinus-Acer-Mercurialis* (W8) in south and *Fraxinus-Sorbus* (W9) in the north.

IRELAND (CROSS 1998): Four sub-types included in the Bohn map, *Quercus petraea* forests with *Hyacinthoides non-scripta* (equivalent to Blechno-Quercetum coryletosum); *Quercus petraea-Fraxinus excelsior* forests with *Corylus avellana, Circaea lutetiana, Brachypodium sylvaticum* and *Veronica montana* (equivalent to Corylo-Fraxinetum veronicetosum and typicum); *Corylus avellana-Fraxinus excelsior* forests on shallow calcareous soils with *Sesleria caerulea* and *Asplenium trichomanes*, rich in bryophytes (equivalent to Corylo-Fraxinetum neckeretosum); and *Corylus-Fraxinus* scrub alternating with *Sesleria* grasslands and *Dryas octopetala* heaths over limestone.

CORINE: 3.1.1

Key factors of forest biodiversity

STRUCTURAL: Ash-beech woods typically occur on freely-drained calcareous soils on sloping sites. Beech is characteristically accompanied by ash, wild cherry *Prunus avium*, field maple, wych elm *Ulmus glabra*, pedunculate oak *Quercus robur*, sessile oak *Quercus petraea*, and locally small-leaved lime and large-leaved lime *Tilia platyphyllos*. Where beech is not absolutely dominant and the soil is shallow, a diverse understorey of yew, holly, whitebeam and other shrubs develops. On the deeper soils, hazel *Corylus avellana* and hawthorn *Crataegus monogyna* appear. Sycamore is well-established in many woods. Dog's mercury *Mercurialis perennis* commonly carpets the ground.

COMPOSITIONAL: Ash-oak woods are dominated by the two species, with hazel as the commonest understorey species. Pedunculate oak is characteristic, but sessile oak occurs on a variety of sites, from strongly acid, poorly-drained clays, to light, acid loams. Field maple, wych elm, wild cherry and sallows *Salix caprea* are frequent in the former; with silver birch *Betula pendula*, small-leaved lime, hornbeam *Carpinus betulus* and alder *Alnus glutinosa* often present in the latter. Hawthorn, dogwood *Cornus sanguinea*, spindle *Euonymus europaeus* and other shrubs are frequently found in the understorey. There is, however, considerable variation in stand composition within these types. Although ash-oak-hazel mixtures are commonest, woods dominated by hornbeam, small-leaved lime, field maple, wych elm, sessile oak or alder are all found. In many woods the stand is a complex, small-scale patchwork of different dominants. On the alkaline or neutral soils (many of which are heavier and more poorly-drained), dog's mercury is the dominant groundflora species; whereas bluebell *Hyacinthoides non-scripta* is the dominant species in those woods on acid soils ranging from poorly-drained clays to lighter, base-poor sandy loams.

NATURAL DISTURBANCES: Major factors of disturbance are grazing by sheep (e.g. in the uplands of U.K.) as well as the prevention of natural regeneration of trees and shrubs due to deer grazing and by small mammals (e.g. grey squirrels) and flooding in riparian areas.

Current land use/silviculture (including history and trends)

Managed as *Tilia* or *Fraxinus* coppice woods. Fragmented and cleared for agricultural use. Afforestation, usually with *Picea abies* in lowland areas. Restoration through thinning and premature felling to remove conifers.

Ashwood F2, photo: Peter Friis Møller.

Ashwood F2

Mixed oak-hornbeam forest F3

Characteristics, ecological conditions and main tree species

This forest is dominated by oak *Quercus* spp. and hornbeam *Carpinus betulus*, and is generally present on clay to lime-clay substrates in plain, colline to submountainous levels and in sub-Atlantic to continental climates. Two potentially natural variants exist which are in competition with beech forests: oak-hornbeam forests on wet soil water regimes (predominantly high water tables) or on dry soils (dominating dry phases in the soil water budget). On wet soils, *Quercus robur* dominates, on dry soils *Quercus petraea* prevail. In both cases beech is not strong enough to compete as a result of root damage.

Mixed oak-hornbeam forests substitute beech forests in areas where beech can not grow as a result of special local climatic conditions (sites with frequent frost periods in early spring, basins with temperature inversion) as well as macroclimac areas with too low precipitation rates.

Oak-hornbeam forests represent successional forests of the former mixed oak, dominant in central Europe before the appearance of beech. Since that time this forest type has been modified and adapted to human needs (e.g. using coppice or coppice with standard management). Presently, only special sites are naturally covered by oak-hornbeam forests. It is noteworthy, however, that *Carpinus betulus* was not an element of post-glacial mixed oak forests but re-immigrated only later from the east, in competition with beech immigrating from the south.

MAIN TREE SPECIES: *Quercus robur, Q. petraea, Fraxinus excelsior, Corylus avellana, Carpinus betulus, Crataegus monogyna, Tilia cordata, Acer pseudoplatanus, Fagus sylvatica, Castanea sativa, Crataegus oxyacantha, Acer campestre, Ulmus carpinifolia, Sorbus torminalis, Prunus avium.*

Main forest types according to country schemes

Selected examples of typical associations:

SWITZERLAND: Galio silvatici-Carpinetum, Carpino betuli-Ostryetum, Arunco-Fraxinetum castanetosum, Cruciato glabrae-Quercetum castanosum, Lathyro-Quercetum.

HUNGARY: Carpino-Quercetum subcarpaticum, Carici pilosae-Carpinetum, Querco roburi-Carpinetum.

ITALY: Ornithogalo-Carpinetum, Querco-Ulmetum (Pianura Padana and pre-Alpine areas, small and rare stands important as examples of the original vegetation of the northern lowlands which are now intensively urbanised or cultivated).

SLOVENIA: Vaccinio myrtilli-Carpinetum, Asperulo-Carpinetum, Helleboro nigri-Carpinetum.

AUSTRIA: Asperulo odoratae-Carpinetum, Galio sylvatici-Carpinetum, Carici pilosae-Carpinetum, Helleboro nigri-Carpinetum.

GERMANY: Galio-Carpinetum, Stellario-Carpinetum, Potentillo albae-Querceetum petreae.

DENMARK: oak-hazel.

U.K.: *Quercus-Rubus-Pteridium* (W10).

FRANCE: Stellario Hollosteae-Carpinetum, Pulmonmario montanae-Carpinetum.

BELGIUM: Endymio-Carpinetum-Primulo-Carpinetum (Carpinion betuli).

NETHERLANDS: Carpinion betuli.

CORINE: 3.1.1

Key factors of forest biodiversity

STRUCTURAL: Typical mixed oak-hornbeam high forests managed by intensive long-term silvicultural treatment show a characteristical two-layered structure. Oak builds up the main layer and hornbeam is used as an auxiliary species in the dominated part of the stand. Most of the silvicultural activities during the rotation period of these stands aim to maintain the lead of oak against hornbeam. Therefore strong signs exist that the structure of naturally developed stands of this type is rather different to stands managed even by close-to-nature silvicultural practices.

COMPOSITIONAL: In the areas where coppicing had a long tradition, especially in the transition zone of beech or hornbeam dominated vegetation belts, the percentage of hornbeam and oak have increased due to their ability of sprouting. Through coppicing the treelayer and understorey have been modified in many cases very efficiently. In this forest type the short-living shade-tolerant species hornbeam forms a highly organised life community with the long-living light depending oak. In comparison to mountainous beech dominated forest types remainders of virgin forests are absent. Therefore the complicated natural stand development cycle of mixed hornbeam-oak forests is not clear. Existing strict forest reserves (not to be confused with real virgin forests, which do not exist in this forest type) show a change of species composition from oak to hornbeam (Mayer and Tichy 1979) due to abandoned silvicultural intervention for the benefit of oak.

NATURAL DISTURBANCES: Due to the lack of natural developed stands little is known about natural disturbances. In existing strict forest reserves it seems that natural disturbance takes place tree by tree and creates a pattern of small gaps. Taken into consideration that oak becomes 400–600 yr old, very few individuals of successful regenerated seedlings of oak in this gaps are enough to sustain the existence of oak in this forest type over the very long life cycle of this natural forest type. It is evident, however, that because of excessive game browsing natural regeneration risks to fail even to this small extent necessary for long time survival.

Current land use/silviculture (including history and trends)

Most stands have been or are being transformed into high forests with the silvicultural aim of valuable oak timber production. Regeneration of oak is mostly achieved through planting, as natural regeneration has been suppressed as a result of browsing, grazing and competition from herbaceous vegetation. Silvicultural alternatives are broad-leaved species of high value (e.g. plantations of *Fraxinus excelsior, Acer pseudoplatanus, Prunus avium* on moist or well-drained fertile soils, and pine or Douglas fir plantations on dry soils).

In Italy most of these very restricted stands are protected as they represent remains of the original vegetation of the actually most cultivated and industrialised areas of the country. Those stands, which are not included in a natural reserve, are strongly threatened by expanding settlements, agriculture and forest plantations.

In the U.K. this forest type is managed either as coppice with standards (hazel or lime coppice), grazed wood pasture (oak pollards) or in high forest management with conversion to plantations (often Norway spruce and pine).

Mixed oak-hornbeam forest F3, photo: Georg Frank.

Mixed oak-hornbeam forest F3

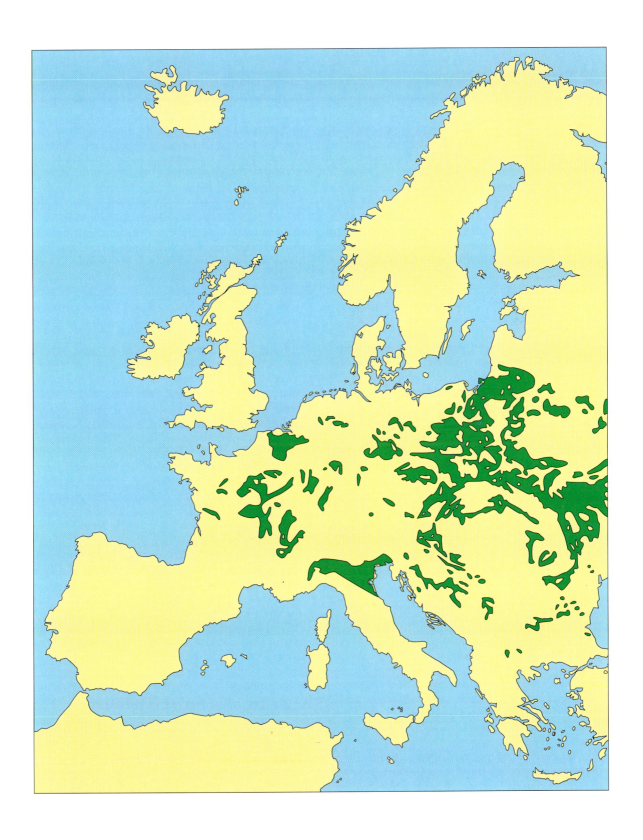

Lowland and submontane beech forest F5a

Characteristics, ecological conditions and main tree species

This forest type is dominated by beech, and is characteristically present on a wide range of soils (e.g. sandy to loamy and acidic to neutral/alkaline soils), a medium water regime, and in areas with Atlantic climates (winters not too cold, summers not too dry, no frost periods in early spring, compare F3). Beech dominates as a result of its competitive strength in monospecific and monolayered stands. Beech forests of the lowlands are probably the most important potential natural forests and showing the most widespread area-proportion of all natural forest types of central Europe.

In the Alps this forest type only occurs up to an elevation of ca 600 m a.s.l. This means that the forest type covers the submontane altitudinal zone of the peripheral Alps but due to less Atlantic but more subcontinental climatic conditions it cannot establish itself in the transition zone and central zone of the Alps. In the southeastern Alps and even in the southwestern Alps this important forest type occurs but more or less as a small altitudinal belt which is not recordable in the map due to the very small and split-up areas. Larger areas are covered by this type in the basins of the southeastern peripheral zone of the Alps. In the low mountain ranges of the Hungarian Great Basin extra-zonal pure beech forests occur. They are restricted to low temperated islands surrounded by zonal oak-forests (see G2).

Beech woodlands in the British Isles have a restricted distribution, being concentrated in what appears to be the native British range of *Fagus* and represented only locally elsewhere by beech plantations (which are often floristically indistinguishable).

MAIN TREE SPECIES: *Acer pseudoplatanus, Fagus sylvatica, Quercus petraea, Q. robur, Fraxinus excelsior.*

Main forest types according to country schemes

Typical examples of selected associations:

GERMANY: Luzulo-Fagetum (+/– or including = Deschampsio-Fagetum, Illici-Fagetum, Milio-Fagetum, Melampyro-Fagetum), Galio odorati-Fagetum (+/–= Melico-Fagetum) Hordelymo-Fagetum (= Lathyro-Fagetum), Carici Fagetum (including Seslerio-Fagetum).

ITALY: Cardamine pentaphyllae-Fagetum, Carici albae-Fagetum, Luzula albae-Fagetum, Veronico-Fagetum.

SLOVENIA: Hacquetio-Fagetum, Hedero Fagetum Luzulo-Fagetum, Blechno-Fagetum, Vicio oroboidi-Fagetum, Hacquetio-Fagetum.

AUSTRIA: Helleboro-Fagetum, Asperulo-Fagetum, Luzulo-Fagetum, Carici albae-Fagetum, Carici pilosae-Fagetum, Lathyro-Fagetum, Aro maculati-Fagetum.

SWITZERLAND: Luzulo silvaticae-Fagetum, Luzula niveae-Fagetum, Streptopo-Fagetum, Galio odorati-Fagetum, Milio-Fagetum, Pulmunario-Fagetum, Aro-Fagetum, Cardamino-Fagetum, Carici albae-Fagetum, Seslerio-Fagetum, Taxo-Fagetum.

IRELAND: *Fagus sylvatica*, occurs in species poor *Quercus petraea* and in *Q. petraea-Fraxinus* excelsior forests.

U.K.: *Fagus sylvatica* occurs in three community types: *Fagus-Mercurialis* woodland, *Fagus-Rubus* woodland, and *Fagus-Deschampsia* woodland, reflecting a shift from calcicolous to more acidic soil profiles.

HUNGARY: Melitti-Fagetum, Daphno laureolae-Fagetum, Melica uniflorae-Fagetum, (Luzulo-Querco-Fagetum).

FRANCE: Mixtures of *Fagus sylvatica* and *Quercus robur* and *Q. petraea*. Classified by French as mixed beech-oak forests and under Corine codes as 41.1 (41.12, 41.13, 41.16). Syntaxon: Ilici fagenion.

DENMARK: Beech wood on limestone on mor (acid soils) and on mull (rich, mainly calcareous soils). Floristic beech-forest types are (Nygaard et al. 1999): *Stachys sylvatica-Geum urbanum/ Fraxinus*-beech forest, *Stachys sylvatica-Oxalis/ Fraxinus* beech-forest, *Galium odoratum-Deschampsia caespitosa/ Corylus* beech forest, *Galium odoratum-Lamiastrum galeobdolon/ Corylus* beech forest, *Galium odoratum-Hepatica nobilis/ Corylus* beech forest, *Deschampsia flexuosa-Carex pilulifera/ Sorbus aucuparia*-beech forest, *Deschampsia flexuosa-Vaccinium myrtillus/ Sorbus aucuparia*-beech forest, *Allium ursinum/ Fraxinus*-beech forest.

BELGIUM: Millet grass-Beech Forest Milio-Fagetum, wood melick- Beech Forest Melico-Fagetum, Calcareous Beech Forest Carici-Fagetum, acidic beech forest with wood rush Luzulo-Fagetum and Atlantic Beech Forest Endymio-Fagetum.

THE NETHERLANDS: Quercion robori-ptraeae p.p., Luzulo-Fagion and Eu-Fagion.

CORINE: 3.1.1

Key factors of forest biodiversity

STRUCTURAL: The natural forest cycle is estimated to be 200–250 yr. This stand age is seldom accepted in commercial forestry because of the outbreak of diseases, the die-back of older trees and a general decrease in timber value. That is why commercial forests are poor in logs

Lowland and submontane beech forest F5a, photo: Peter Friis Møller.

and snags. This managed pure beech forests usually have the tendency to develop a dense closed canopy and only one layer. As analysis of structure in reminders of virgin forests (Mayer 1971, Korpel 1995) shows that under natural or seminatural conditions stands are much more structured and even multi-layered stand types can occur.

COMPOSITIONAL: Because of the limited light supply as a result of the closed canopy the understorey is built up by very shade-tolerant species and in most sub-types shrub species are absent. Only on shallow and poorer soils (*Quercus* spp., *Acer pseudoplatanus*), on soils with a better water regime, and better base saturation (*Fraxinus excelsior*), on drier sites (*Quercus* spp., *Acer* spp.) and in successional phases (*Acer pseudoplatanus, Betula* spp., *Fraxinus excelsior*) can other species temporarily interfere. In general, no or only few shrubs occur (e.g. *Corylus avellana, Sambucus nigra, S. racemosa,* and *Lonicera* spp.) because of shade conditions in the stands. In beech woods in the British Isles, shade-tolerant species like *Ilex aquifolium* and, to a lesser extent, *Taxus baccata,* tend to increase over other woody associates. Ground vegetation varies from very poor (acidic soils) to very rich (fresh, calcareous and base-rich sites). In the southern Alps almost pure beech forests are present also on thermophilous sites.

NATURAL DISTURBANCES: Apart from catastrophic windthrow main processes of forest development is restricted by small gap dynamics, but frequently casual gaps close, so that also uneven aged stands show closed canopies and appear homogeneously structured, with one layer, dark, and poor in shrubs and even ground vegetation. Beech is rather sensitive to the parasitic fungus *Fomes fomentarius*, which often is an important factor in gap-formation.

Current land use/silviculture (including history and trends)

This forest type is typically managed as high forest, with a 120–150 yr rotation time, often with use of natural or seminatural regeneration. In Germany many stands are converted into oak-beech or oak-hornbeam coppice with standards. Compared with other tree species, beech immigrated very late into the northern parts of central Europe after the latest ice age, often around 1500–500 BC. Beech has not been able to settle on all of its potential sites as a result of anthropogenic forest utilisation. Although, in many lowland areas, e.g. Denmark, beech has been favoured for mast production until 1800 AD and since around 1800 AD by drainage and other silvicultural activities. Many potential beech forest areas were transformed to spruce plantations, frequently after agricultural use.

In Italy most of these very restricted stands are protected as they represent remnants of the original vegetation of the most cultivated and industrialised areas of the country. Those stands, which are not included in natural reserves, are strongly threatened by the expanding settlements, agriculture and forest plantations.

In the U.K., beech stands are managed under high forest systems, although some were formerly managed under selection systems, e.g. in the Chilterns. Management is now on a clear-fell or shelterwood system. Many are established in mixtures with other species, often conifers. Where these have been planted on ancient woodland sites, current policy is to restore the sites to semi-natural conditions, where practicable. Beech woods previously managed as coppice-with-standards or as wood-pasture are of particular value for biodiversity. Many of the surviving and former wood-pastures are of such high value that their management must be the subject of consultation, and most are Sites of Special Scientific Interest (SSSIs).

Lowland and submontane beech forest F5a

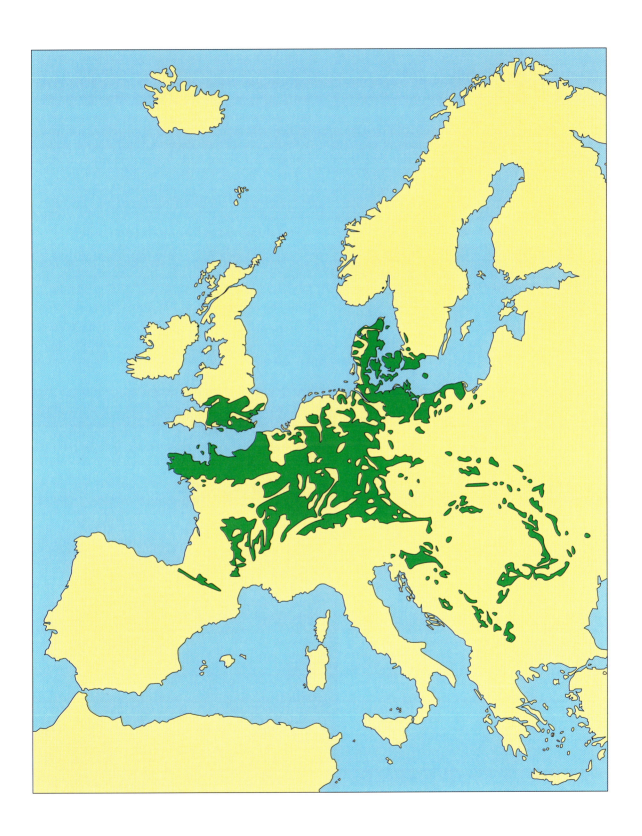

Montane beech and mixed beech-fir-spruce forests F5b

Characteristics, ecological conditions and main tree species

Alps

SOUTHERN ALPS: Landscape structure is very diverse due to orographic factors such as the direction of mountain ranges, exposition, elevation and steep altitudinal gradients. These factors create more differentiated and highly diverse sets of ecological conditions than is noted in the northern Alps.

NORTHERN ALPS: The ecological and climatic conditions are much more uniform than in the southern Alps, therefore the altitudinal and geographical zonation is more distinct. Altitudinal belts are clearly defined into either a submontane (nearly pure beech forests, see F5a) or a broad montane belt (of mixed beech-silver fir-spruce forests).

In both geographical sub-areas, geological substrate (e.g. limestone and silicate bedrock) is the critical factor controlling forest type, (rendzina vs other soil types). This site factor has also become important in the history of human settlement, agricultural land use and forest history: the limestone landscape was not as suitable for agriculture as those landscape types dominated by silicate substrate for example. As a consequence, almost all remnants of central European and alpine virgin forests are found in limestone areas today.

Central European low mountain ranges

The natural beech forests of the Bavarian Forest, Black Forest, Vosges and Swiss Jura are mixed with silver fir, irrespective of the chemical character of the bedrock (silicious as well as calcareous stones). On rocky sites or on sites with

Montane beech and mixed beech-fir-spruce forests F5b, photo: Peter Friis Møller.

mobile stones and blocks, where competitiveness of beech is reduced, sycamore *Acer pseudoplatanus* and mountain elm *Ulmus scabra* may play an additional role. Sycamore is also the most important secondary tree species during regeneration phases (pioneer and intermediate forests).

MAIN TREE SPECIES: *Fagus moesieca, Fagus sylvatica, Abies borissi-regis* (hybrid fir)*, Abies alba, Acer pseudoplatanus, Quercus cerris, Quercus pubescens, Ilex aquifolium, Taxus baccata, Pinus sylvestris, Pinus leucodermis, Pinus nigra.*

Main forest types according to country schemes

The described foresttype appears mainly in the Alps: Switzerland, France, Austria, Germany, Slovenia, Italy. They are ecologically and structurally similar despite being described in the phytocoenological literature using different nomenclature and exist in the climatically exposed mountainous regions of Central Europe (Bavarian Forest and, less evident, also in the northern and western low mountain range).

Selected typical examples of associations:
ITALY: Cardamine trifolia-Fagetum, Oxalido-Fagetum.
SLOVENIA: Omphalodo-Fagetum (Abieti-Fagetum), Lamio orvalae-Fagetum, Anemono trifoliae-Fagetum, Homogyne sylvestris-Fagetum, Polysticho lonchitis-Fagetum.
AUSTRIA: Helleboro nigri (Abieti)-Fagetum, Asperulo odoratae-(Abieti)-Fagetum, Luzulo nemorosae-Fagetum, Aposerido-Fagetum, Anemono-Fagetum, Aceri-Fagetum, Lamio orvalae-Fagetum, Calamagrostio villosae-Fagetum, Dentario enneaphylli-Fagetum.
GERMANY: Galio odorati-Fagetum (including former Abieti-Fagetum), Aceri-Fagetum, Lonicerae alpigenae-Fagetum (including Dentario heptaphylli-Fagetum), Cardamino trifoliae-Fagetum, Luzulo-Fagetum.
SWITZERLAND: Abieti-Fagetum typicum, Abieti-Fagetum luzuletosum, Abieti-Fagetum polystichetosum, Aceri-Fagetum.
FRANCE: Dentario pentaphyllo-Fagetum, Buxo-Fagetum, Abieti-Fagetum luzuletosum niveae, Abieti-Fagetum luzuletosum sylvaticae.
SPAIN: Scillo lilio-hyacinthi-Fagetum, Buxo-Abieti-Fagetum, Luzulo nivae-Fagetum, Avenello-Fagetum.
CORINE: 3.1.3

Key factors of forest biodiversity

STRUCTURAL: A sparse shrub and herbal layer is characteristical for pure beech and mixed beech stands which is a difference from other colline and submontane forest types with dominating broad-leaved species. The main difference in structure between the lowland beech forest (F5a) and mixed beech-fir-spruce forests (F5b) is as follows: in the Abieti-Fagetum the canopy architecture is much rougher because the conifers may distinctly rise above the crowns of the beech.

COMPOSITIONAL: This forest type has rather distinct plant-geographical gradients (west–east, south–north) that exist beside the altitudinal gradient, with local variation and overlap according to microclimatic and geological site conditions. Differences between the northern and southern Alps: along a gradient from the dinaric region over the peripheral alpine region and further to the central alpine region, the percentage of spruce increases. A similar gradient exists from the west to the east in the southern Alps. The spruce percentage increases towards the east whilst the silver fir percentage increases towards the west. Differences between the alpine and dinaric region: in the Dinaric mountains beech is more dominant in high altidudinal regions. Special sub-types also exist, where spruce and fir are not competitive. Beech may represent the upper timberline here.

NATURAL DISTURBANCES: Large-scale disturbances do not occur in this forest type, due to the high diversity in composition and structure of these stands. Natural disturbances create small gaps in the order of magnitude of one tree-length in diameter.

Current land use/silviculture (including history and trends)

In general, agriculture has had a less significant impact on limestone area habitats than the mining industry (including salt mining). In comparison to the central Alps there is an outer zone ("peripheral Alps") where limestone is dominant, and therefore forest cover has not been changed as much by agriculture. Large differences between the northern and southern Alps, in terms of forest cover and forest characteristics, are caused by the impact of very intensive cultivation of the pre-Christian northern Italian culture, and of the Venetian Renaissance. It is therefore necessary to differentiate between the northern, southern and south-eastern (including Dinaride) geographical regions.

In the south-eastern Alps, regular forest management started very late (in some remote regions of the Dinarides and Bosnia as late as in the 19th century) during the period of exploitation by the forest administration of the Austrian-Hungarian Monarchy. Also the human influence has changed the species composition a lot, supporting the dispersal of the spruce and the fir and pushing back the beech.

Montane beech and mixed beech-fir-spruce forests F5b

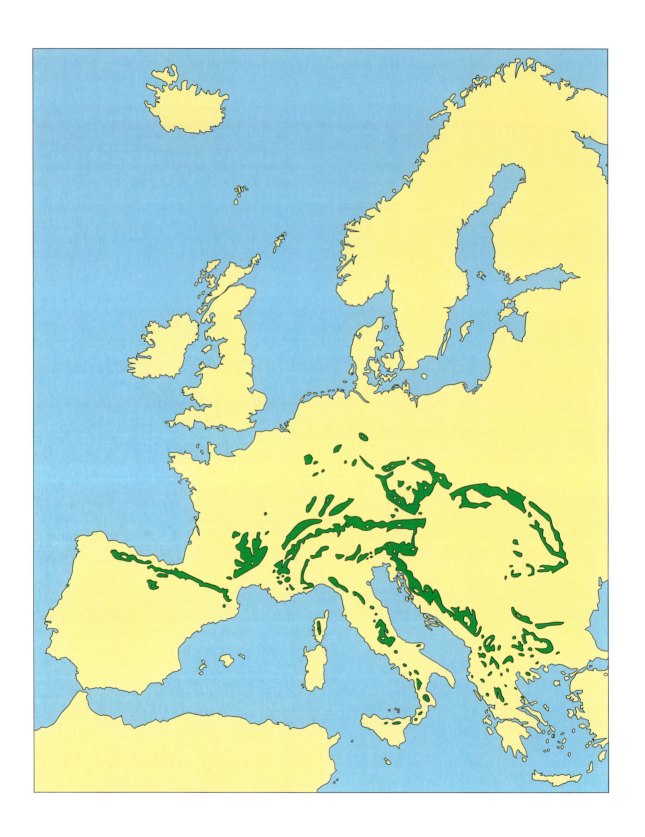

Mediterranean and submediterranean mixed oak forest G2+G3

Characteristics, ecological conditions and main tree species

These forest types appear mainly in the southern European region of mixed broad-leaved forests. In the Continental biogeographic region they are present in the mainly Pannonian basins and lowlands, in the area surrounded by the Carpathians in the north-east, the Balkan mountains in the south-west and the Alps in the north-west. It is further important in Greece, Italy, Spain, France, in south-west and south of Germany and Portugal, and it exists on the slopes of main Atlantic rivers.

Soils vary from average brown soils to dry rendzina-types. In many cases the geological substrate consists of loess or sand dunes with corresponding dry soils.

MAIN TREE SPECIES: *Acer monspessulanum, Acer obtusatus, Carpinus orientalis, Carpinus betulus, Castanea sativa, Fraxinus ornus, Ostrya carpinifolia, Pistacia terebinthus, Quercus coccifera, Q. petraea, Q. cerris, Q. pubescens, Q. conferta, Q. robur, Q. pyrenaica, Sorbus torminalis, Tilia* spp.

Main forest types according to country schemes

SLOVENIA: Querco pubescentis-Carpinetum orientalis, Querco-Ostryetum carpinifoliae, Seslerio autumnalis-Ostryetum.

AUSTRIA: Quercetum petraeae-cerris, Aceri tatarici-Quercetum, Corno quercetum pubescentis, Geranguinei-Quercetum pubescentis, Potentillo albae-Quercetum.

SWITZERLAND: Quercetea robori-petraeae.

GERMANY: Quercetum pubescenti-petraeae, Potentillo albae-Quercetum petraeae.

GREECE: Quercetum Cocciferae, Coccifero Carpinetum, Carpinetum Orientalis, Quercion Confertae, Tilio Castanetum, Quercetum Montanum, Ostryo-Carpinoum.

ITALY: Quercion pubescentis types, Orno-Ostryenion types, Quercion frainetto types, Oleo-Ceratonion type.

SPAIN: *Quercus pubescens* forests, *Q. pyrenaica* forests, *Q. faginea* forests, *Q. canariensis* forests, chestnut forests, *Betula* spp.

Mediterranean and submediterranean mixed oak forest G2+G3, photo: Peter Friis Møller.

PORTUGAL: *Quercus pyrenaica* forests, *Q. faginea* forests, *Q. canariensis* forests, chestnut forests, *Betula* spp. forests.

FRANCE: Thermophile oak forest, on calcicole to neutrophile substrates, mainly made of *Quercus pubescens* and *Q. robur* (Quercetalia pubescenti-petraea).

CORINE: 3.1.1

Key factors of forest biodiversity

STRUCTURAL: Little is known about the structure and natural dynamics of these types because only a few relicts of forests of this type exist due to long-term human impact. Due to a high number of tree and shrub species occurring in this type, multi-layered structures and complex patchiness appear to be characteristical. Key factors include fragmentation and patchiness, continuity and connectivity of forest cover (ecological corridors), ecotones, structure, vertical distribution of vegetation and ground layers, forest regeneration, small specific habitats (bark and decayed wood), litter (quality and amount).

COMPOSITIONAL: In the continental biogeographic region these forests are probably the species richest because of its multi-layered and very complex structure. Elements of Mediterranean, Continental and even Atlantic flora regions overlap in these forests. Species with specific stand type and scale requirements, such as large herbivores, or raptors are also an important key factor.

NATURAL DISTURBANCES: Natural disturbance regimes and development cycles of this forest type are not well studied due to the lack of remainders of natural forests of this type. Natural disturbances include wildfire, insect pests, fungal disease and browsing.

Human impact has been severe, with intensive coppicing/logging, overgrazing that had influences on this forest type. At the other end of the scale, the abandonment of agricultural areas enables re-colonisation. Possible effects of invading non-native species on ecological processes (allelopathy or nitrogen fixation).

Current land use/silviculture (including history and trends)

For > 5000 yr man has removed the forests in the oak-dominated lowlands. Very large areas are therefore completely deforested. Remainders of forest land could only survive on terrains or soils less suitable for agriculture. Today forest re-colonisation processes can be observed in abandoned agricultural land and grassland, in Greece, Spain and Portugal. Some areas of this forest type are protected by national conservation laws (in National and Natural parks). In France, these forests are commonly managed by private owners for multiple uses (coppicing, grazing, fuelwood supply, agroforestry). In Italy, forestry aims also at restoring degraded forest (e.g. coppice conversion, thinning, afforestation).

Mediterranean and submediterranean mixed oak forest G2+G3, photo: Susana Dias.

Mediterranean and submediterranean mixed oak forest G2+G3

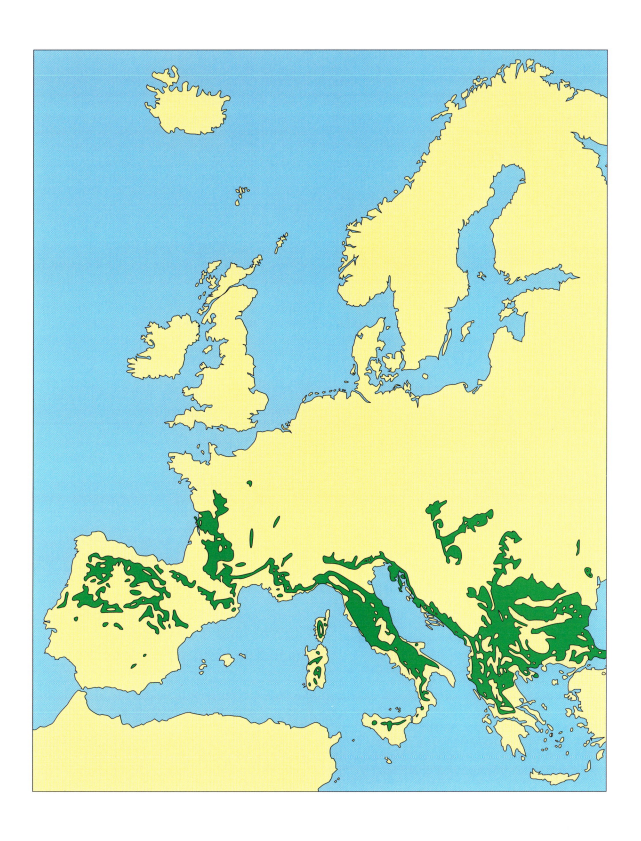

Mediterranean broad-leaved sclerophyllous forest and shrub J1–J8

Characteristics, ecological conditions and main tree species

Very typical in southern and coastal Greece/Italy (including islands, mainly Sardinia, Corsica and Sicily), widespread in Spain but less so in Portugal.

This type of forest is the dominant type of vegetation in a transition zone between temperate forest, mostly dominated by deciduous trees and the scrublands (maquis, chaparral, etc), that herald the subtropical regions. Plants have to cope with a selective pressure resulting from a double stress – winter cold and summer drought – which has determined their morphological and ecophysiological evolutionary responses. In most parts of the Iberian Peninsula, holm oak is largely indifferent to soil reaction whereas in the northern areas it prefers calcareous soils and southern slopes where it competes successfully with deciduous species (Terradas 1999).

In Portugal and Spain this type leads to the "montados" and "dehesas" where scattered trees (mainly oaks) are integrated in traditional agro-forestry systems.

GREECE: *Ceratonia siliqua, Olea europea* var. *sylvestris, Quercus ilex, Q. coccifera.*

ITALY: *Quercus ilex, Q. suber, Q. coccifera, Olea europea* var. *sylvestris, Ceratonia siliqua.*

SPAIN: *Quercus rotundifolia, Q. suber, Q. ilex, Q. coccifera, Olea europea* var. *sylvestris, Ceratonia siliqua, Chamaerops humilis, Maytenus europeus.*

PORTUGAL: *Quercus rotundifolia, Q. suber, Olea europea* var. *sylvestris, Ceratonia siliqua, Quercus coccifera.*

MACARONESIA: *Chamaemeles coriacea, Dracaena draco, Maytenus umbellata, Olea europaea* spp. *maderensis* and spp. *cerasiformis, Pistacia atlantica, P. lentiscus, Phoenix canariensis, Sideroxylon mirmulanus.*

Main forest types according to country schemes

GREECE: Oleo-Ceratonietum, Oleo-lentiscetum, Adrachnotus-Quercetum ilicis, Orno-Quercetum ilicis.

ITALY: Quercion ilicis types, Oleo-Ceratonion types.

SPAIN: *Quercus rotundifolia* forests, *Q. suber* forests, *Q. ilex* forests, *Olea europea* forests and woodlands, *Ceratonia siliqua* woodlands, *Q. coccifera* shrublands, *Chamaerops humilis* bushlands, *Maytenus europeus* scrubs.

PORTUGAL: *Quercus rotundifolia* forests, *Q. suber* forests, *Olea europea* forests and woodlands, *Ceratonia siliqua* woodlands, *Q. coccifera* forest and shrublands.

CORINE: 3.2.3/3.2.4

Key factors of forest biodiversity

STRUCTURE: Age of trees, mosaic patterns, habitat fragmentation, continuity of forest cover, stand structural complexity, forest regeneration, fuel accumulation and litter accumulation are considered to be of major importance as key factors; the type of tree species in agro-systems have moderate implications for biodiversity, because it does not imply significant differences in structure. The fact that these types of forest have roughly diffuse edges, makes it difficult to establish the real importance of edges characteristics to biodiversity.

COMPOSITIONAL: Stand type specific species, namely the understorey composition, as well as the biological soil conditions are crucial for this forest type.

NATURAL DISTURBANCES: Wildfire could, in some areas, be of significant importance for this forest type, but to a much lesser extent than for pine forests or plantations. Drought is a major factor in this region, leading to the presence of species with particular adaptations to this stress, like the evergreen holm oak, which is much more resistant to drought than winter-deciduous trees but not enough to survive in the drier areas of the Mediterranean. Extreme drought can produce high rates of crown withering and tree mortality. High losses of hydraulic conductivity can occur during summer drought and losses can be much larger than in other Mediterranean trees (*Ceratonia siliqua, Quercus suber,* and *Olea europea*). Drought could be also responsible for interrupting the closed canopy, promoting the scarcity of trees and restriction to best sites, with subsequent implications on the structure and functioning of the ecosystem. Insect pests and browsing are also very important. Human influences: Forest protection status, forest practices (intensive coppicing on *Quercus ilex, Q. rotundifolia, Q. coccifera*; branching, cork extraction), soil erosion, agricultural pressure, grazing pressure and social influences are all of major importance.

Current land use/silviculture (including history and trends)

There are great geographical differences in management of this forest type. On very oligotrophic soils in the plains of southern Iberian Peninsula, holm oak and cork oak woodlands are managed as "dehesas" (in Spain) and "montados" (in Portugal), a man-made type of savannah-like ecosystem with large, isolated trees emerging from a grassland. These and related systems in Sardinia, Greece and elsewhere combine extensive grazing of natural pastures with intermittent cereal cultivation (oats, barley and wheat) in park-like woodlands of cork oak and holm oak. These formations result both from selection and protection of superior well-shaped trees occurring among natural stands and in some regions from the intentional planting of acorns chosen from selected trees. Tree density is maintained at 20–40 ha^{-1} and mature trees are pruned regularly to remove infested branches, broaden their canopy cover and increase acorn production (Joffre et al. 1988 in Blondel and Aronson 1999). Various products are also collected during the process, including timber, firewood, charcoal, tannin and natural cork. Non-wood goods like mushrooms, honey, and game species are also important. Although theses systems are of anthropogenic origin, they show remarkable stability, biodiversity and sustained productivity as a result of their balanced two-purpose vegetation structure, heavy incorporation of animal husbandry and botanically rich mosaic-like herbaceous plant layers (Joffre and Rambal 1993 in Blondel and Aronson 1999).

In other regions like eastern Spain and in most of the Mediterranean regions of Italy, Greece and France closed-canopy holm oak forest are prevalent and charcoal and firewood were the main products obtained from it. As a result these forests have been managed as coppices or coppice-like standards at least since the neolithic period. Coppicing has created forests with high tree densities, and the ensuing intense competition leads to a low growth rate of individual stems. Charcoal production, very important in the past, has practically been abandoned. Tannin from the bark of young trees was another major product but synthetic tannins have replaced such production in the last 30 yr. The main current use is firewood. Multiple use is in fact occurring but not as a result of planning. Coppice has been always accompanied by sheep and goat grazing. Acorns play an important role in the diet of sheep during winter (Terradas 1999).

The major threats were, and to some extent are, the replacement by fast growing pines or large stands of *Eucalyptus* for timber and pulp production, increasing deforestation and semi-industrial clearing in order to extend mechanised cropping lands, and the intense use of pesticides and fertilisers. Due to recent EU policy, trends are shifting to the recognition of economically viable multi-use of this agro-systems.

In areas, where less or no management occurs, this forest type can exhibit a high degree of complexity, promoting rather diverse ecosystems, with relatively high degree of specificity, at least in their floristic composition (Alves et al. 1998), like the vegetation of Barrocal limestone in Algarve (south Portugal). Some of these considered as important sites for conservation under the Portuguese law.

Mediterranean broad-leaved sclerophyllous forest and shrub J1–J8, photo: Peter Friis Møller (upper), Susana Dias (lower).

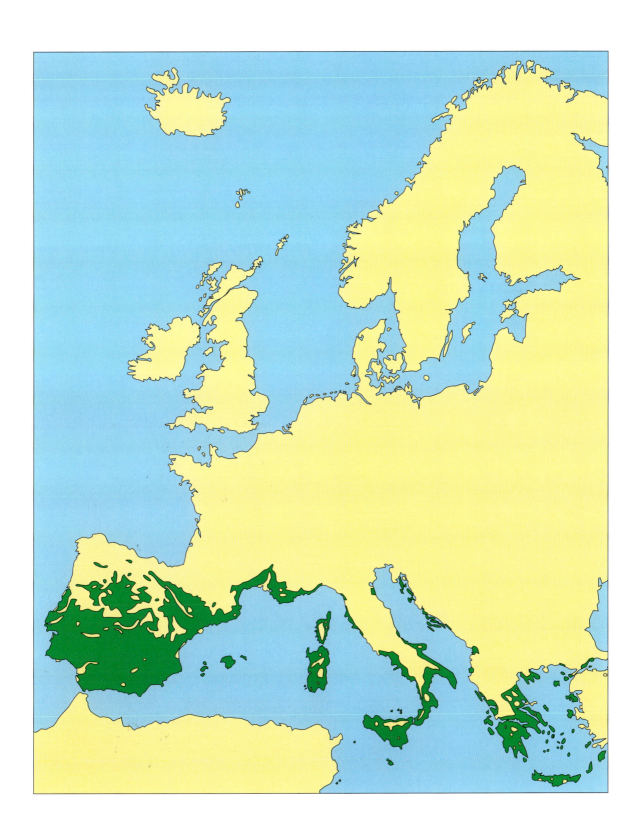

Mediterranean and Macaronesian coniferous forests, woodlands K1–K4

Characteristics, ecological conditions and main tree species

In this unit are included the Oromediterranean and Thermomediterranean coniferous forests and woodlands of the Mediterranean biogeographical region.

The Oromediterranean forests are typical of the mountainous areas of southern Greece and Spain and of southern Italian Appennines. These forests are pioneer formations of black pine, white barked pine, cypress or Mediterranean firs; such communities grow mainly on poorly developed soils on steep and rocky slopes. Oceanic conditions and siliceous soils allow the development of mesogean pine formations (in Spain up to 1300 m, in Tyrennian coasts of Italy up to 700 m). *Pinus pinea, Pinus halepensis* and *Pinus brutia* formations are Thermomediterranean pioneer community, largely widespread in all Circummediterranean coastal areas.

The typical physiognomy of such types is an open cover of coniferous species, namely in pine formations, allowing the development of a more or less closed scrub layer.

Edaphic constraint is a major key factor of forest dynamics. Wildfire might allow the persistence of a pioneer community, by triggering off the process of forest regeneration in mesogenean pine and *Pinus halepensis* forests.

GREECE: *Pinus nigra, Abies cephalonica, Pinus brutia, Pinus halepensis, Cupressus sempervirens, Pinus pinea, Pinus sylvestris, Pinus leucodermis, Juniperus* spp.

ITALY: *Pinus nigra, Pinus laricio, Pinus pinaster, Pinus halepensis, Pinus pinea, Pinus leucodermis, Juniperus* ssp.

SPAIN: *Pinus pinaster, Pinus halepensis, Pinus sylvestris, Pinus nigra, Pinus pinea, Juniperus* spp.*, Abies pinsapo, Tetraclinis articulata.*

PORTUGAL: *Pinus sylvestris, P. pinaster, P. pinea* and *Juniperus* spp. The dimensions of these types of forests in Portugal have no significant dimension, but this same species are now widespread by plantations.

MACARONESIA (IN CANARIES): *Pinus canariensis, Juniperus cedrus, Juniperus turbinata* spp. *canariensis.*

Mediterranean and Macaronesian coniferous forests, woodlands K1–K4, photo: Peter Friis Møller.

Main forest types according to country schemes

GREECE: Acero-Cupression, Pinetum nigrae (*Cupressus sempervirens* forests, *Pinus leucodermis* forests, *Pinus sylvestris* forests, *Juniperus* spp. forests, *Pinus halepensis* forests, *Pinus brutia* forests, *Pinus pinea* forests).

ITALY: Orno-Ericion types (20–21), Rumici-Astragalion type, Pino-Genistum aspalathoidis types, Euphorbio ligusticae Pinetum Pinastri type.

SPAIN: *Pinus pinaster* forests, *Pinus halepensis* forests, *Pinus sylvestris* forests, *Pinus nigra* forests, *Pinus pinea* forests, *Juniperus* spp. woodlands, *Abies pinsapo* forests, *Tetraclinis articulata* woodlands.

CORINE: 3.1.2

Key factors of forest biodiversity

STRUCTURAL: Mosaic patterns, habitat fragmentation, continuity of forest cover, stand structural complexity, forest regeneration, fuel, dead wood, litter accumulation.

COMPOSITIONAL: Stand type specific species (alien species). In Italy *Pinus* spp. types are dominated by these trees, once introduced by forest plantations and today naturalised. Wildfire, major insect pests, fungal disease, browsing, atmospheric pollution are other key factors.

NATURAL DISTURBANCES: The most important key factor is wildfire. Human influences: Agriculture and grazing pressure, tourism and recreation pressure, and coastal urbanisation. Logging pressure (selective cutting and shelterwood cutting). Existence of designated conservation areas (protected forest reserves and special management regimes).

Current land use/silviculture (including history and trends)

Wood production, fuel supply (juniper woodlands), resin extraction, grazing, honey production (*P. halepensis*, *P. brutia*), soil protection, water balance, tourism and recreation.

Mediterranean and Macaronesian coniferous forests, woodlands K1–K4

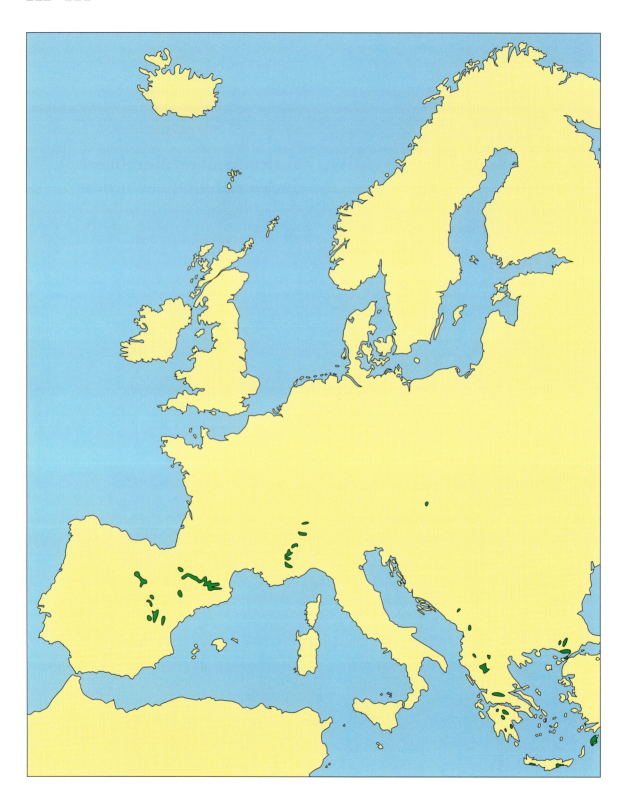

Atlantic dune forest P1

Characteristics, ecological conditions and main tree species

Small area in Europe: on siliceous substrates, in the dune, of southwestern France, ca 100 000 ha, in the western part of Portugal, and very restricted areas in the coastal border of north Netherlands, western Denmark and Germany. Great importance for genetic resources for pine production in France. In the southern Atlantic littoral, there are three main species: maritime pine *Pinus pinaster*, holly oak *Quercus ilex* and cork oak *Quercus suber*. Also *Pinus pinea* and *Populaus tremula* can occur. In the northern littoral the main species are *Quercus robur* and *Betula pendula*.

Main forest types according to country schemes

Pure pine or mixed (with oaks) high forests. The two main vegetation types in France are Pino maritimi-Quercetum suberis and Pino maritimi-Quercetum ilicis and in Germany Populo tremulae petraeae-Quercetum. They are established on poorly transformed soils with wind deposits, with a moder-mor humus (pH 4–5). The associated flora is xerophile to mesoxerophile and oligotroph.
CORINE: 3.1.3

Key factors of forest biodiversity

STRUCTURAL: Topography is the determinant of abiotic conditions at a fine scale influencing both soil dynamics and micro-climate, creating heterogeneity in stand composition. Because of even-aged management, age of dominant trees is the main factor influencing both vertical structure of vegetation and canopy closure. Consequently biodiversity must be considered over groups of stands in either different silvicultural phases, or different tree compositions. Moreover, internal edges between stands affect biodiversity.

COMPOSITIONAL: Topography is the determinant of abiotic conditions at a fine scale influencing both soil dynamics and micro-climate, creating heterogeneity in stand composition. Because of even-aged management, age of dominant trees is the main factor influencing both vertical structure of vegetation and canopy closure. Consequently biodiversity must be considered over groups of stands in either different silvicultural phases, or different tree compositions. Moreover, internal edges between stands affect biodiversity.

NATURAL DISTURBANCES: The main natural disturbances are: insect and root diseases damages, fire and windstorms. Recreation management pressure and aerial spraying of insecticides. Reforestation practices.

Current land use/silviculture (including history and trends)

This forest is extensively managed with multi-use objectives: dune protection against erosion, recreation activities, and wood pine production. Insecticides are often sprayed to reduce defoliator impacts.

Atlantic dune forest P1, photo: Peter Friis Møller (upper), Susana Dias (lower).

Atlantic dune forest P1

Ombrotrophic mires S1

Characteristics, ecological conditions and main tree species

This type is a complex of different peatlands with a poor nutrient status. Throughout the boreal region ombrotrophic mires are important components of the landscape. Raised bogs are found in southern Fennoscandia extending into the south boreal zone. The local importance of these mire types is dependent on the topography and regional humidity. Having a flat topography, mires are particularly important in NE Sweden and N Finland. In Norway, where the climate is Atlantic and the topography dramatic, mires form blanket bogs, which do not cover large areas, like in Finland and Sweden. The type is characterised by the dominance of *Pinus sylvestris*, mixed with *Betula pubescens* spp. *pubescens*. Near the Atlantic Sea *Picea abies*, *Betula pubescens* spp. *pubescens* and *Alnus* spp. can be dominating.

Main forest types according to country schemes

The main site types in different Fennoscandian countries range from forested ombrotrophic pine mires (*Sphagnum fuscum*) to treeless aapa mires. Ombrotrophic mires include a moderate variation in nutrient richness and moisture levels, and consequently in ground and field vegetation, but most typically exist on nutrient poor, rather dry paludified sites.
CORINE: 4.1.2

Key factors of forest biodiversity

The most important key factor maintaining the characteristic biodiversity of this type is the water regime of the site. However, due to rather dry ground conditions, and water

Ombrotrophic mires S1, photo: Peter Friis Møller.

supply by rain and not ground water, this forest type may dry out completely during hot summers. As a consequence they do not necessarily form no-fire refugia, and fire can be regarded as main natural disturbance factor in this forest biodiversity type.

STRUCTURAL: Hollows in the peat cover offers a wide range of habitats for different mire species. The typical form of diameter distribution of trees in natural stage is reversed J-shape and tree cover may be quite sparse and the productivity is low.

COMPOSITIONAL: The type is characterised by the dominance of *Pinus sylvestris*, mixed with *Betula pubescens* spp. *pubescens*. Near the Atlantic Sea *Picea abies*, *Betula pubescens* spp. *pubescens* and *Alnus* spp. can be dominating. The ground layer is dominated by a continuous coverage of *Sphagnum* spp. mosses and the field layer is evenly covered with dwarf shrubs (*Vaccinium* spp.) and sedges (e.g. *Carex* spp. and *Eriophorum vaginatum*). Several endangered species of the boreal zone are mire species. Regionally their proportion can be very high. For example, several ombro-oligotrophic mire plant species (e.g. *Carex globularis*, *Rubus chamaemorus*, *Ledum palustre* and *Betula nana*) which are still common in Fennoscandia, are rare or even endangered in other parts of the Europe. Ecotones between mires and forests are diversity hot spots for plants, epiphytic lichens, bracket fungi and butterflies.

NATURAL DISTURBANCE: Under natural conditions in dry summers fire may occur as a tree stock replacing disturbance. However, typically the long continuity and low productivity causes a reversed J-shape diameter distribution of trees. Despite the low productivity of nutrient poor pine mires they have been intensively ditched for forestry use, especially in Finland. Ditching naturally changes the soil conditions and consequently reverts ground and field vegetation towards forest vegetation. Ditching also causes a growth effect, which changes the natural structure of forest stock, further changed by the consequent forestry measures.

Current land use/silviculture (including history and trends)

The 2/3 of the total mire area is ditched in Finland. However, no further ditching is made today and due to their large area coverage, ombrotrophic mire complexes are well represented in existing conservation areas.

Ombrotrophic mires S1

The type occurs all over south and middle Fennoscandia, only a few larger mires are indicated on the map.

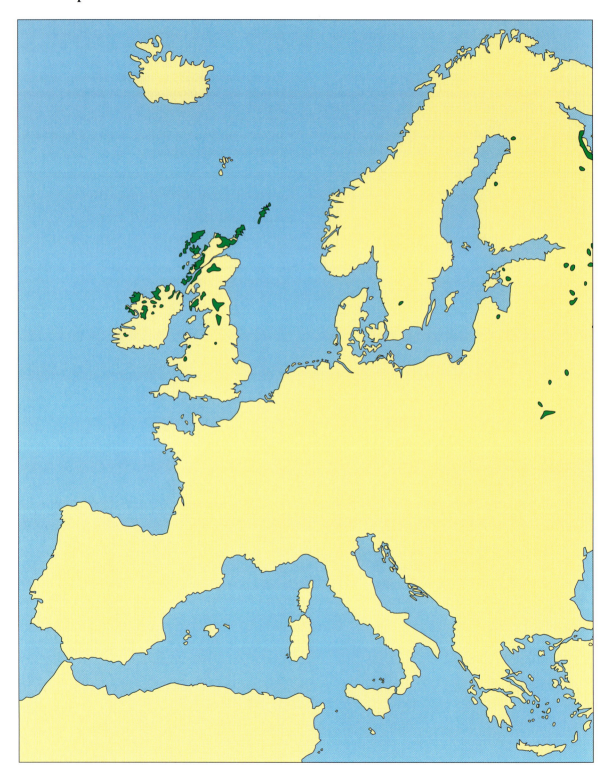

Minerotrophic mires incl. swamp forest S3

Characteristics, ecological conditions and main tree species

This type is an important component throughout the boreal zones in Fennoscandia and covers a wide range of forest habitats. The tree cover varies as well as the relative proportion of *Pinus sylvestris* and *Picea abies*. The nutrient richness varies from moderate to rich and moisture level is high. The ground layer is characterised by sedges (*Carex* spp.), some grasses, and tall herbs.

Main forest types according to country schemes

The forest biodiversity type covers a very wide range of mire types used in Finland (see further subdivision of forest types). In Sweden and Norway, where there is no separate classification system of mire types, the nutrient richness of this type ranges from moderate to rich and moisture level is high.
Corine: 4.1.2

Key factors of forest biodiversity

The most important key factor maintaining the characteristic biodiversity of this type is the water regime of the site. Due to a continuous ground water supply this forest type usually forms no-fire refugia. It can also be characterised by a quite large amount of coarse woody debris, especially on the most fertile sites.

Structural: The hollows in the peat cover offer a wide range of habitats for different mire species. In natural conditions the forest type has a wide variation in diameter distribution and the typical form of diameter distribution in natural stage is reversed J-shape. In the most fertile sites the coarse woody debris is an important structure.

Compositional: The composition of the tree species and ground vegetation varies a lot according to the subdivision of the forest biodiversity type (see Further subdivision of forest types). Many species of nesting birds fauna are totally, or in some phases of their life cycle, dependent on these mire types. Also the ecotones between mires and forests are diversity hot spots for plants, epiphytic lichens, bracket fungi and butterflies.

Natural disturbances: Under natural conditions the main disturbances are small-scale gap-dynamics due to lethal frost or fungal diseases. Due to the high productivity of minerotrophic mires the natural dynamics of this type has been intensively affected by ditching and forestry measures in Finland and Sweden.

Current land use/silviculture (including history and trends)

At present no new areas are ditched and on the most valuable mire areas restoration of the natural conditions is even carried out by covering the ditches. New protection areas of these mire complexes have been established during the last decade, but the problem is the low area proportion of these mire types of the protected areas, especially in the southern boreal region.

Further subdivision of forest types

In the following the type is divided into pine and spruce dominated mires of three different nutrient status classes.

Pine mires, sedge-level (poor) nutrient status: Characterised by *Pinus sylvestris* dominance. Productivity is low, forest stock can be sparse. A sparse mixed tree layer of *Betula pubescens* spp. *pubescens* might be abundant. Ground layer is characterised by the occurrence of sedges (*Carex chordorrhiza, C. lasiocarpa, C. rostrata*) and *Sphagnum papillosum*. The field layer is quite evenly covered with dwarf shrubs.

Spruce mires, sedge-level (poor) nutrient status: Wet mires. Characterised by poorly developed *Picea abies* and *Betula pubescens* spp. *pubescens* coverage. Ground layer dominated by *Carex* genera (*C. echinata* very typical). Low coverage of grasses.

Pine mires, herb-rich (medium) level: Wet mires with varying thickness of peat layer. Dominated by poorly developed *Pinus sylvestris* coverage. Almost always a mixture of *Betula pubescens* spp. *pubescens* exists. Dwarf shrubs can exist on the dryest spots, the sedge coverage is diverse and herbs (*Potentilla palustris, Menyanthes trifoliata*) also exist.

Spruce mires, herb-rich (medium) level: Characterised by poorly developed *Picea abies, Betula pubescens* spp. *pubescens* mixture (sometimes birch dominated). Moss layer (*S. angustifolium*) is continuous, and tall-sedges (like *C. chordorrhiza*) are typical. Herb vegetation is quite diverse (*Potentilla palustris, Menyanthes trifoliata, Equisetum palustre*).

Pine mires, eutrophic (rich) level: Tree coverage *Pinus sylvestris* dominated, but *Betula pubescens* spp. *pubescens, Alnus* spp. and also *Picea abies* may exist. Moss and sedge coverages continuous. A lot of grasses, and *Menyanthes trifoliata, Equisetum palustre*. Exists only in regions with alkaline base rock.

SPRUCE MIRES, EUTROPHIC (RICH) LEVEL: Thin peatlayer, characterised by dense spruce (*Picea abies*) coverage. Hardwood species may exist in small groups. Moss coverage is continuous and herb vegetation is very diverse (e.g. *Solidago virgaurea, Saussurea alpina, Potentilla erec-*

ta). Also *Juniperus communis* is typical. This mire type exists almost only in some regions near the border of the middle and north boreal zones.

The main site types in different Fennoscandian countries varies from forested pine and spruce mires to treeless fens.

Minerotropic mires incl. swamp forest S3, photo: Peter Friis Møller.

Minerotrophic mires incl. swamp forest S3

The type occurs all over north and middle Fennoscandia, only a few larger areas are indicated on the map.

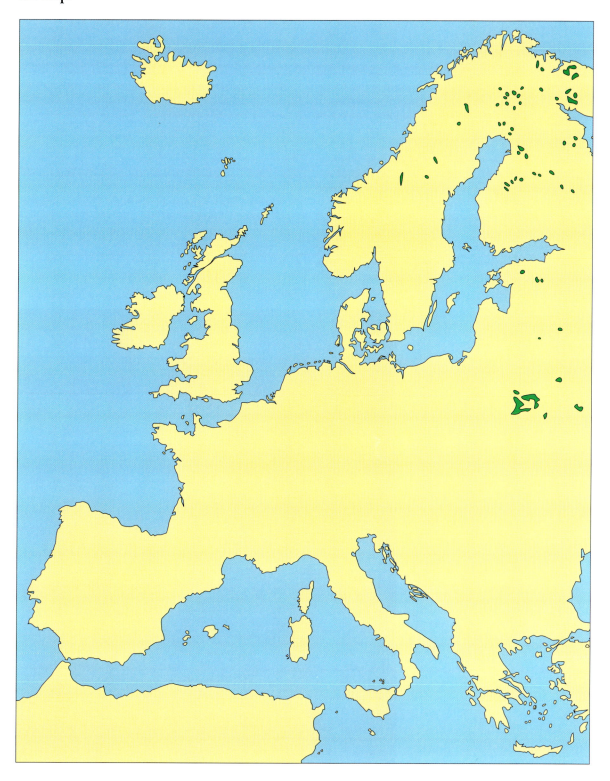

Swamp and fen forests, alder T1

Characteristics, ecological conditions and main tree species

HABITAT: Wet to waterlogged organic soils, base-rich and moderately eutrophic; fen peats in open water transitions, flood-plain mires with influences of ground water, often under influence of (winter) flooding. Important (dominant) species:

TREE LAYER: *Alnus glutinosa, Betula pubescens, Quercus robur, Fraxinus excelsior*. In the Alps this vegetation occurs only in few sites, in some no-reclaimed alpine valleys where the stagnant water has not been regimented yet.

SHRUB LAYER: *Salix cinerea, Sorbus aucuparia, Frangula alnus*.

HERB LAYER: *Carex paniculata, Calamagrostis canescens, Iris pseudacorus, Scutellaria galericulata, Solanum dulcamara*.

MOSS LAYER: *Sphagnum palustre, S. squarrosum, Calliergonella cuspidata*.

Main forest types according to country schemes

GREAT BRITAIN: (W1 *Salix cinerea-Galium palustre* woodland) (W3 *Salix pentandra-Carex rostrata* woodland) W5 *Alnus glutinosa-Carex paniculata* woodland W6 *Alnus glutinosa-Urtica dioica* woodland.

BELGIUM: Carici laevigatae-Alnetum Carici elongatae-Alnetum Carici remotae-Alnetum Cirsio-Alnetum.

NETHERLANDS: Alnion glutinosae.

GERMANY: Carici laevigatae-Alnetum Carivi elongatae-Alnetum Sphagno-Alnetum.

SLOVENIA: Carici remotae-Fraxinetum, Carici elongatae-Alnetosum glutinosae.

AUSTRIA: Stellario nemorum-Alnetum glutinosae, Carici elongatae-Alnetosum glutinosae.

ITALY: Carici elongatae-Alnetosum glutinosae.

SWITZERLAND: Carici elongatae-Alnetosum glutinosae.

CORINE: 4.1.1

Swamp and fen forests, alder T1, photo: Peter Friis Møller.

Key factors of forest biodiversity

Base status maintained by ground-water influence or periodic flooding; uprootings are frequent and create canopy gaps, mosaic structures and niches for many plants and animals (e.g. amphibians).

Current land use/silviculture (including history and trends)

Remnants that escaped clearing and cultivation into grassland have little economical value; nature reserves, sometimes formerly coppiced. This forest type was probably very common in the northern plains of Italy, where it disappeared as a result of cultivation and industrialisation. Now it occurs only in some alpine valleys, threatened by the works done to regulate natural water courses, and it has to be considered very rare.

Swamp and fen forests, alder T1, photo: Peter Friis Møller.

Swamp and fen forests, alder T1 and swamp and fen forest, birch T2

Also in northern Italy, cf. text.

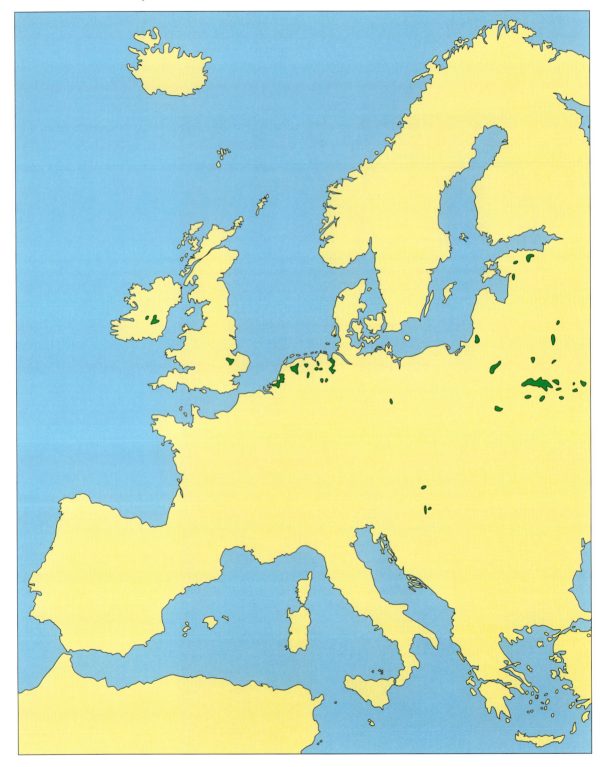

Swamp and fen forests, birch T2

Characteristics, ecological conditions and main tree species

HABITAT: Moderately acid, nutrient poor peaty soils in a variety of mire types; often in association with bogs and wet heaths; not influenced by base rich ground water nor flooded; more influenced by rain water than ground water; susceptible to nutrient input (ground water, atmospheric deposition).

TREE LAYER: *Betula pubescens, B. pendula, Pinus sylvestris*. In Ireland: *Salix cinerea, Alnus glutinosa, Fraxinus excelsior* and *Betula pubsecens* (areas influenced by ground water); or *Betula pubescens, Ilex aquifolium, Sorbus aucuparia* and *Quercus robur* (more oligotrophic and drier sites) (Cross 1998).

SHRUB LAYER: *Salix aurita, S. cinerea.*

HERB LAYER: *Molinia caerulea, Dryopteris dilatata, Vaccinium* spp.

MOSS LAYER: *Sphagnum* spp., *Polytrichum commune*. In Ireland: *Rubus fruticosus, Hedera helix, Pteridium aquilinum, Dryopteris dilatata, Molinia caerulea* (with remnants of original bog surface dominated by *Calluna vulgaris*) (Cross 1998).

Main forest types according to country schemes

U.K.: (W2 *Salix cinerea-Betula pubescens-Phragmites australis* woodland), W4 *Betula pubescens-Molinia caerulea* woodland.

BELGIUM: Vaccinio-Betuletum pubescentis.

NETHERLANDS: Betulion pubescentis.

GERMANY: Betuletum pubescentis (Vaccinio uliginosi-Pinetum sylvestris) Betuletum carpaticae.

IRELAND: A subset of Irish PNV type "Degraded raised bogs with *Alnus* carrs, *Fraxinus-Alnus* and *Betula* forests, fens and heaths" (Cross 1998).

CORINE: 4.1.1

Key factors of forest biodiversity

STRUCTURAL: Zonations and mosaics with other types; successional phase in birch invasion of mires.

NATURAL DISTURBANCE: Due to rather wet ground conditions, this forest type forms a fire refuge.

Swamp and fen forests, birch T2, photo: Peter Friis Møller.

Current land use/silviculture (including history and trends)

No economical value. Often drained and forested with line plantations; nature reserves. In Ireland, this forest type is part of "a mosaic of vegetation communities which occur in complex on degraded, i.e. cut-away and drained, raised bogs, and they reflect the very varied substrata which arise following peat extraction"(Cross 1998).

Swamp and fen forests, alder T1 and swamp and fen forest, birch T2

Flood plain (alluvial and riverine) forests U1+U2

Characteristics, ecological conditions and main tree species

Important (dominant) species in tree layer: *Fraxinus excelsior*, *Ulmus minor*, *Acer pseudoplatanus*. In Ireland, *Quercus robur*, *Fraxinus excelsior*, *Alnus glutinosa*, *Betula pubescens*. Shrub layer: in Ireland, *Corylus avellana*, *Crataegus monogyna* and species of shrub willows (e.g. *Salix cinerea* ssp. *oleifolia*). Herb layer: in Ireland: *Filipendula ulmaria*, *Circaea lutetiana*, *Primula vulgaris*, *Hedera helix*, *Rubus fruticosus*, *Urtica dioica*, *Carex remota*, *Deschampsia caespitosa*, *Geum rivale* (Cross 1998).

Main forest types according to country schemes

U.K.: W7 *Alnus glutinosa-Fraxinus excelsior-Lysimachia nemorum* woodland.

BELGIUM: Carici remotae-Fraxinetum Primulo-Fraxinetum excelsioris Stellario-Alnetum Filipendulo-Alnetum Violo odoratae-Alnetum Ulmo-Fraxinetum Salicetum triandro-viminalis.

NETHERLANDS: Alno-Padion.

GERMANY: Stellario nemorum-Alnetum glutinosae Carici remotae-Fraxinetum Pruno padi-Fraxinetum Chrysosplenio oppositifolii-Alnetum glutinosae.

IRELAND: Wetland woodland (Heritage Council 2000). Alternative scheme: Alluvial forests with *Quercus robur*, *Fraxinus excelsior* and *Salix* species, and *Alnus-Quercus-Corylus* forests with *Salix cinerea* ssp. *oleifolia*.

CORINE: 3.1.3

Flood plain (alluvial and riverine) forests U1+U2, photo: Konstantinos Spanos.

Current land use/silviculture (including history and trends)

Parks open to public, poplar plantation, grassland, hardly any spontaneous natural forest.

Further subdivision of forest types

Regularly flooded and never flooded (polders). Ireland: alluvial forests with *Quercus robur*, *Fraxinus excelsior* and *Salix* species (forest unit 8 of Cross 1998) occur on alluvium in the flood plains of river valleys. *Alnus-Quercus-Corylus* forests with *Salix cinerea* ssp. *oleifolia* occur on heavy, poorly drained gleyed clays – not subject to flooding, but which are wet in winter and dry out on the surface in summer.

Flood plain (alluvial and riverine) forests U1+U2

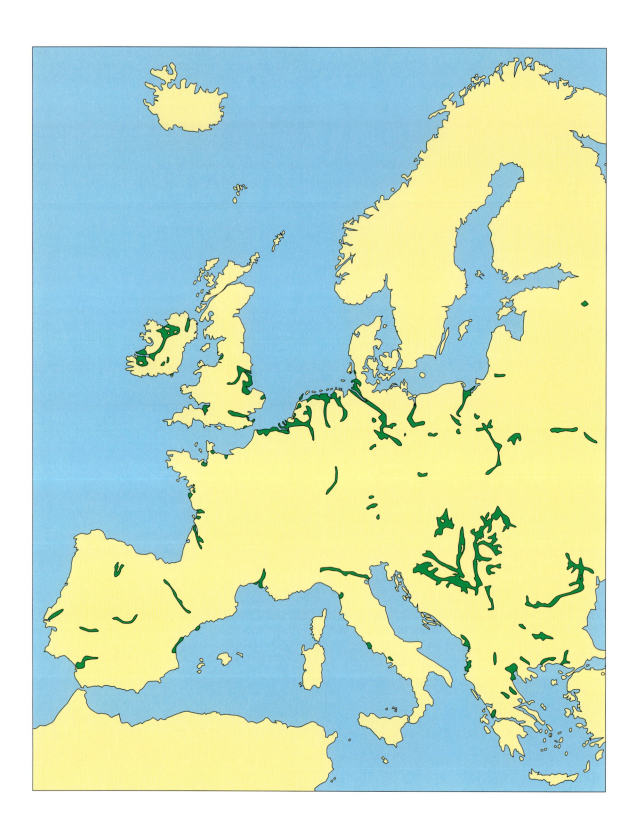

Mediterranean and Macaronesian riverine woodlands and gallery forests U4

Characteristics, ecological conditions and main tree species

Flood plain and riverine forests occupy minor areas in Mediterranean countries but include many riverine habitats sheltering relevant components of biodiversity and playing important roles for biological conservation issues (e.g. presence of rare species), for landscape connectivity maintenance (e.g. ecological corridors) and for protective functions. The geomorphological traits of the Macaronesian archipelagoes reduce even more the stretch of this forest type, but the same considerations about its importance can be applied.

MAIN TREE SPECIES:

GREECE: *Platanus orientalis, Alnus glutinosa, Salix* spp., *Populus* spp.

ITALY: *Carpinus betulus, Laurus nobilis, Alnus cordata, Alnus glutinosa, Salix elegans, Salix purpurea, Nerium oleander, Platanus orientalis.*

SPAIN: *Fraxinus angustifolia, Ulmus minor, Alnus glutinosa, Populus* spp., *Salix* spp., *Tamarix* spp., *Nerium oleander, Flueggea tinctoria.*

PORTUGAL: *Fraxinus angustifolia, Ulmus minor, Alnus glutinosa, Populus* spp., *Salix* spp., *Tamarix* spp., *Nerium oleander, Flueggea tinctoria.*

MACARONESIA: *Salix canariensis, Tamarix* spp., *Phoenix canariensis, Populus* spp.

Main forest types according to country schemes

GREECE: *Platanus orientalis* forests, *Alnus glutinosa* forests, *Populus* spp. forests, *Salix* spp. forests

ITALY: Laurel-oak woodland, *Alnus cordata* woods, Tyrrhenian ash-alder galleries, oleander *Nerium oleander* galleries, purple willow scrub, Sicilian plane tree canyons.

SPAIN: Mixed flood plain forests with *Fraxinus angustifolia*, mixed flood plain forests with *Ulmus minor, Alnus glutinosa* riparian forests, *Populus* spp. riparian forests, *Salix* spp. riparian forests and scrub, tamarisk *Tamarix* spp. galleries, oleander *Nerium oleander* and "tamujo" *Flueggea tinctoria* galleries.

PORTUGAL: Mixed flood plain forests with *Fraxinus angustifolia*, mixed flood plain forests with *Ulmus minor, Alnus glutinosa* riparian forests, *Populus* spp. riparian forests, *Salix* spp. riparian forests and scrub, tamarisk *Tamarix* spp. galleries, oleander *Nerium oleander* and "tamujo" *Flueggea tinctoria* galleries.

MACARONESIA: Canarian willow galleries *Salix canariensis*, tamarisk galleries *Tamarix* spp., palm groves of *Phoenix canariensis, Populus* spp. riparian forests.

CORINE: 3.1.3

Mediterranean and Macaronesian riverine woodlands and gallery forests U4, photo: Susana Dias.

Key factors of forest biodiversity

Geomorphology of river-beds and basins, soil properties and the zonation determined by periodic flooding and periodic variations of ground-water level are general factors influencing the structure and diversity of riverine forests and woodlands.

STRUCTURAL: The intrinsic distribution patterns of these forests favour the importance of factors related to area, habitat fragmentation (connectivity, edges) and continuity of forest cover. Other relevant factors are: stand structural complexity, distribution of gaps generated by flooding, patterns of forest regeneration, dead wood and litter. The naturalness of water courses is also relevant at landscape level.

COMPOSITIONAL: Stand type specific species and alien species (riverine habitats are prone to invasions by alien species).

NATURAL DISTURBANCES: The natural disturbance regimes are marked by flooding dynamics. Browsing and fungal or insect pests can also be relevant. Management of river-beds and water courses (hydraulic regulation, drainage) are of main importance, along with silvicultural practices, interactions with agriculture and grazing and social influences. Physical and chemical pollution (due to agrochemicals and industries) are potentially important.

Current land use/silviculture (including history and trends)

Flood plain soils are among the most productive of Mediterranean countries, and since ancient times have been largely deforestated for agriculture and meadows. The limited water resources of Mediterranean areas have also promoted hydraulic regulation of rivers affecting their natural flooding regimes. Agricultural activities have induced in some areas drainage of wetlands and overexploitation and depletion of aquifers. These trends are hard to reverse in areas of intensive agriculture. Coppicing and branching are silvicultural practices relevant in some subdivisions and areas. Forest type management presently focuses on the maintenance and enhancement of protective functions (notably flood prevention) and conservation-restoration of natural riverine habitats (see poplar plantations).

Laurel forest

Characteristics, ecological conditions and main tree species

This very characteristical and diverse forest type occurs only in the Atlantic islands, with relict occurrences in southern Portugal. It includes all the vegetation formations of the Azores and Canaries and also the most hombrofilic part of Madeira. This forest is a partial relict of a previous type, which covered most of the southern part of Europe and NW Africa during the Tertiary era. The dramatic climatic changes that occurred in the Pleistocene pushed the limits of this forest type southwards into the archipelagoes of Azores, Madeira and Canaries. It hosts a great biological richness, with many relict endemic taxa, but also with several others having a more recent origin, some of them have undergone a pronounced process of speciation within these islands. This forest type is essential for captivation and retention of water, edaphogenesis, and protection against erosion, and plays an important role in the maintenance of other ecosystems integrity processes.

The climatic and habitat diversity within and between the different archipelagoes supports the existence of different types of laurel forests, under humid and sub-humid conditions of the Azores, Madeira and Canaries. These differences are enhanced by the evolutionary history of each one of these archipelagoes, as the Macaronesian laurel forest is a sui generis consequence of occasional immigration phenomena. The biogeographical links of Macaronesia are very diverse. Despite this, many of the nearest relatives are in the Mediterranean-North Anatolian-Caucasian-Iranian Himalayan belt, e.g. *Laurus azorica*, *Prunus lusitanica* and *Pistacia atlantica*. However, there is also a very old element in the Macaronesian flora with disjunct distribution of families and genera that suggest their origin in the pre-drift landmass of Gondwanaland, e.g. Lauraceae, Myricaeae and *Clethra*.

MADEIRA: *Ocotea foetens, Laurus azorica, Myrica faya, Clethra arborea, Persea indica, Apollonias barbujana, Ilex perado, Ilex canariensis, Picconia excelsa, Heberdenia excelsa, Pittosporum coriaceum, Visnea mocanera.*

CANARIES: *Laurus azorica, Persea indica, Ocotea foetens, Apollonias barbujana, Ilex perado, Picconia excelsa, Myrica faya, Heberdenia excelsa, Ilex canariensis, Myrica rivas-martinezii, Prunus lusitanica* ssp. *hixa, Visnea mocanera, Arbutus canariensis, Myrsine canariensis.*

AZORES: *Frangula azorica, Juniperus brevifolia, Laurus azorica, Piconia azorica, Myrica faya, Vaccinium* spp.

Laurel forest, photo: Francisco Manuel Fernandes.

Main forest types according to country schemes

AZORES: *Dracaena draco* formations, *Myrica faya* and *Erica* formations, humid laurel forest (Pruno-Lauretalia azorica), Hyper-humid laurel forest (*Vaccinum* spp.), *Ilex* forest, *Juniperus* forest (Juniperon brevifoliae).
CORINE: 3.1.1

Key factors of forest biodiversity

STRUCTURAL: The most important factors are related to tree species, mosaic pattern, habitat fragmentation, contuinity of forest cover (ecological corridors), stand structural complexity, forest regeneration, dead wood and litter.
COMPOSITIONAL: The tree species and understorey composition, herbivorous and the edaphic conditions, soil (pH), water in the soil (soaked areas) also plays a major role as key factors.
NATURAL DISTURBANCES: Wind, snow and drought are considered as having slight to moderate importance as key factor. Human influences: forest practices, grazing pressure, of minor importance, whereas forest protection status are considered to be important for the maintenance of this forest type.

Current land use/silviculture (including history and trends)

The exploitation for timber, fuel, agriculture and grazing in this forest began in the 15th century with strong decrease of its natural distribution. Nowadays it is mainly considered as protected forest in most of the area, being included in Natural Reserves or other conservation schemes (e.g. Natura 2000, being also considered world heritage by UNESCO); no special silviculture use (timber, fuel are not important nowadays). In the Canaries and Madeira there are some areas which are periodically exploited for local agricultural purposes.

Further subdivision of forest types

MADEIRA: Clethro-laurion. These are woodlands dominated by *Erica scoparia* ssp. *maderinicola, Ilex canariensis, Vaccinium padifolium, Myrica faya*, not yet described (Fernandes et al. pers. comm.).
AZORES: The general type can be subdivided in three main groups according mostly to the humidity conditions:
  a) Thermo Macaronesica (*Dracaena draco* formations and *Myrica faya* and *Erica* formations);
  b) Meso Macaronesica – laurel forest (mainly with *Laurus* spp., *Prunus* spp.;
  c) Supra Macaronesica (humid laurel forest, mainly with *Vaccinium* spp., wet *Ilex* forest and *Juniperus* forest).
CANARIES: Ixantho-Laurion azoricae (laurel woodlands pluristratified and great floristic diversity; fayo-Ericion arborae (degraded laurel woodlands); woodlands of *Erica scoparia* var. *platycodon* and *Ilex canariensis* at high windy places.

Laurel forest, photo: Peter Friis Møller.

Hedgerow

Characteristics, ecological conditions and main tree species

Hedgerows (tree- or scrub-lined field boundaries) provide an important biodiversity resource in otherwise poorly forested landscapes (in particular at low altitudinal agricultural land), particularly in Britain and Ireland. Strong correlations between length/area of hedgerow and abundance/diversity of breeding birds, in particular, are well established (e.g. Lack 1992). The contribution of hedgerows to biodiversity is probably higher in Ireland than in other European countries, given the low forest cover (<10%, excluding hedgerows and scrub). Small (1995) recorded a mean of 5.7 km of hedgerow or treeline per 1 km² square in the Ireland (based on 729 systematically distributed squares), or the equivalent of 1.4 ha km⁻². Inclusion of hedgerows as a forest type will provide an opportunity for standardisation of management and monitoring of hedgerows as a biodiversity resource. Inadequate consideration of hedgerow as a "real" habitat type (rather than simply a by-product of archaic agricultural systems) has probably contributed to the ongoing loss of this habitat. In Ireland, the main tree species are ash *Fraxinus excelsior*, hawthorn *Crataegus monogyna*, blackthorn *Prunus spinosa*, willows *Salix* spp., with others including birch *Betula* and oak *Quercus*.

Main forest types according to country schemes

In a draft classification of Irish habitats, hedgerows are currently (provisionally) placed under "Miscellaneous: boundary enclosure; hedgerow" (Heritage Council 1999).

Current land use/silviculture (including history and trends)

Hedgerows and boundary treelines are associated with lowland agriculture, in particular pasture. Management is relatively haphazard and often negative: trimming of hedgerows to maintain "stock-proof" (or simply "tidy") field boundaries and "safe" roadside verges, and hedgerow removal to increase field-size for easier machine access or economies of scale. Hedgerow losses in Britain and Ireland have accelerated rapidly since the mid-century and are widely believed to have contributed to reductions in populations (and to a lesser extent diversity) of farmland birds (e.g. O'Connor and Shrubb 1986). Reviewing British, Swiss and Finnish studies, Lack (1992) concluded that "to retain a high density of birds and a broad range of species the density of hedges (on lowland farms) should not be allowed to fall below ca 60–80 m ha⁻¹, which means an average field size of ca 4–7 ha."

Hedgerow, photo: Peter Friis Møller.

Chestnut coppice

Characteristics, ecological conditions and main tree species

This forest type is present mostly in the Southern Alps (Italy, Switzerland, partly Austria and Slovenia, at altitudes from 300–400 to 1000–1200 m a.s.l.) and very rarely in the lowlands of central Europe (Germany, Switzerland, Austria). It is characterised by the presence of *Castanea sativa*, a species introduced and encouraged by man mainly for fruit production. The chestnut tree is an introduced species but not dominant under natural conditions. The actual presence of pure chestnut coppices is due entirely to the human influence. This forest type exists on very different types of soil, from poor to mesic or fertile moist soils, mainly of siliceous origin. Other important tree species include *Fraxinus excelsior*, *Fagus sylvatica*, *Betula pendula*, *Tilia cordata*, *Acer pseudoplatanus*, *Quercus* spp. (see also Cost Action G4).

Main forest types according to country schemes

ITALY: Querco-Castanetum, Querco-Fraxinetum.

AUSTRIA: This type is not included in the national system due to its agro-forestry origin.

FRANCE: Quercion robori-petraeae, Luzulo-Fagion, Quercion pubescenti-petraeae. Mainly in Corse, Cevennes, Ardèche, Dordogne and Limousin.

SWITZERLAND: e.g. Phyteumno betonicifoliae-Quercetum castanosum.

GERMANY: e.g. Luzulo-fagetum, colline + submontainous Galio-Fagetum.

Key factors of forest biodiversity

STRUCTURAL: The structure and composition of these woods depend strictly on how intensive the human influence has been, and on the previous and present use of these forests. Conditions differ between pure coppice stands used mainly for timber production and chestnut plantations for fruit production. The decline in importance of both these products, and the effects of the bark cancer *Cryphonectria parasitica* have significantly changed the structure and composition of chestnut stands. Where once fruit and litter production were the main goals, old and very big chestnut trees are now the last remnants of original plantations. Presently a lot of other tree species, mainly pioneer species such as *Betula pendula*, *Populus tremula*, *Corylus avellana*, and *Rubus* spp., are replacing the chestnut trees and making these stands develop into more natural vegetation types. As a consequence of this process, the old traditional chestnut plantations are disappearing and/or are in bad condition. Previously intensively coppiced stands, which once were used for timber production, are now becoming older and beginning to suffer from regeneration and stability problems.

COMPOSITIONAL: The species composition of these stands has been changing during the last years. In some regions, chestnut is decreasing and is being replaced by other species.

NATURAL DISTURBANCES: The most important natural disturbance to affect this forest type in recent decades has been "bark cancer". This disease caused the disappearance of traditional forms of agro-forestry in many areas of the Alpine region.

Current land use/silviculture (including history and trends)

Traditional agro-forestry practices, regeneration opportunities, and very old trees left from previous agro-forestry are now being incorporated into stands because of the expansion of forests in areas previously occupied by fruit plantations. Traditional silvicultural management of chestnut coppice stands is also being abandoned and transformed into new management practices. In Austria this agro-forestry type is a historical remnant which only occurs at very few places with both particularly acidic soil conditions, wine-growing climate and a specific history (settlements of Romans and specific agro-forestry types of the late Middle Age, probably combined with abandoned vineyards where *Castanea sativa* was used for fruit production but also for wine-sticks).

Main forest types according to country schemes

Within Britain, there is no classification scheme that adequately describes plantation stands comprised of non-native species. However, using the National Vegetation Classification, it is possible to assess the extent to which a stand relates to a semi-natural woodland analogue. *Picea abies* has been planted on a range of site types in lowland Britain, either as a single species or in mixtures (frequently with oak). Often, this has replaced ancient semi-natural woodland on heavier clay soils, such as W8 – *Fraxinus-Acer-Mercurialis* woodland; or W10 – *Quercus-Pteridium-Rubus* woodland.

Picea sitchensis, in contrast, has often been used in afforestation of upland moorland, covering a range of plant community types: U4 – *Festuca ovina-Agrostis capillaris-Galium saxatile* grassland; U5 – *Nardus stricta-Galium saxatile* grassland; U6 – *Juncus squarrosus-Festuca ovina* grassland; M18 – *Erica tetralix-Sphagnum papillosum* raised and blanket mire; M19 – *Calluna vulgaris-Eriophorum vaginatum* blanket mire; M20 – *Eriophorum vaginatum* blanket

and raised mire; H12 – *Calluna-Vaccinium* heath; H10 – *Calluna-Erica* heath; H21 – *Calluna-Vaccinium-Sphagnum* heath.

It has also replaced some upland, semi-natural woodland community types, such as W17 – *Quercus petraea-Betula pubescens-Dicranum majus* woodland, towards the North and West of Britain.

In Continental Europe no strong tradition exists to classify plantations in the stricter sense because the research approach traditionally focused on natural or semi-natural forests. In general a distinction is made between secondary conifer forests "Sekundärer Nadelwald" and conifer plantation "Nadelholz-Forst" (Hofmann 1994, Starlinger 2000).

Key factors of forest biodiversity

STRUCTURAL: Spruce plantations are (on a stand scale) typically even-aged and mono-layered (with little or no shrub layer), and with a restricted diameter-distribution. There is usually only a minor amount of deadwood (es-

Spruce plantation, photo: Winfried Bücking.

pecially large pieces), a closed canopy, and no clustering of stems. However, manipulation or encouragement of increased structural diversity, relating to such factors, can provide an opportunity to have a marked positive influence on biodiversity at stand and (perhaps more particularly) landscape scales.

COMPOSITIONAL: Spruce plantations frequently comprise or are dominated by a single tree species, with little or no shrub layer in many cases. However, where soil and other conditions allow, the addition (or encouragement) of additional tree species (particularly broad-leaved species, can have a marked influence on biodiversity. The extent of the shrub layer can also be influenced by changes in the light regime, e.g. through increased thinning or through modification of edges. As with structural key factors, manipulation of the composition of spruce plantations can potentially have marked benefits for forest biodiversity.

NATURAL DISTURBANCES: The vast majority of existing spruce plantations are primarily managed for timber production, with consequent implications for biodiversity. There is an increasing tendency to modify such management in ways that will minimise negative influences, and maximise positive influences, on biodiversity, while at the same time maintaining commercial viability. The relative homogeneity of spruce (and other) plantations, compared with natural or semi-natural forest, increases the risk of serious damage by insect pests, wind, snow and ice. Such factors can have negative impacts in both economic and biodiversity terms, although some impacts (e.g. creation of deadwood "snags") can, on a limited scale, be positive for biodiversity.

Current land use/silviculture (including history and trends)

As noted, spruce plantations primarily have a commercial purpose, but there is increasing opportunity and willingness to allow "trade-offs" with biodiversity. For example, with the publication of the U.K. and Irish forestry standards, there is greater emphasis placed on management of forests (even "commercial" ones) to meet multiple purposes. Consequently, there are numerous adjustments being recommended to conventional silvicultural practice. These include encouraging greater proportions of broad-leaved tree species (it has been recognised, for example, that encouraging birch in upland spruce plantations can bring multiple benefits for biodiversity), making increased use of natural regeneration when regenerating forests, ensuring that a minimum amount of open space is retained or created within the forest, increasing the amount of deadwood that is retained within the stand, allowing a proportion of the stand to grow on beyond economic felling age, as long-term retentions.

Many of the upland plantations of *P. sitchensis* in the U.K. have reached the stage where they are ready for felling and replanting, providing opportunities for restructuring. The Forestry Commission in the U.K. has developed a process of landscape evaluation known as Forest Design, and there has been great emphasis placed upon this process in the reshaping of these single-species forests. Felling coupes have been modified in order to produce more organic shapes, which fit the landform, also producing a more varied age structure across the forest at the landscape scale.

In continental Europe as a result of human drift to the cities, abundant farm land was afforested with Norway spruce plantations. Sitka spruce was not used in Continental Europe.

Further subdivision of forest types

Spruce plantations range from monotypic stands through mixed conifer stands (e.g. Sitka spruce with lodgepole pine *Pinus contorta*) and, to a lesser extent, mixed conifer/broad-leaved plantations.

Spruce plantation, photo: Peter Friis Møller.

Poplar plantation

Characteristics, ecological conditions and main tree species

Poplar plantations have been included in the forest types list, although the origin and features of these stands make them much more similar to agricultural land use types. The poplar clones are regularly placed and planted using typical agricultural practices. No character allows a distinction between ordinary agriculture and these kind of agro-industrial landuse except the longer rotation period of ca 10–30 yr, depending on nutrient and climatic conditions. Even natural regeneration, seen as one of the fundamental prerequisites of ecological sustainability, is not possible. They are managed through clear-felling and have to be artificially regenerated. Poplar plantations occupy areas previously colonised by flood plain vegetation, riparian trees and shrubs, being a part of the agricultural landscape.

Main forest types according to country schemes

Different forest types cannot be distinguished because of the totally artificial origin of these plantations.

Key factors of forest biodiversity

STRUCTURAL: Because of the artificial origin, the structure of the popular plantations is pronounced monotonous and even-aged on the stand scale. In every case these plantations have replaced the species-rich and complex structured natural riparian or flood plain vegetation.

COMPOSITIONAL: The totally artificial one species or one-clone composition and the cultural practices (use of chemicals, short rotation) make these stands not much suitable for the life of wild animals and plants.

Poplar plantation, photo: Konstantinos Spanos.

NATURAL DISTURBANCES: There is more or less no influence of natural disturbances on the dynamic of these plantations, which are totally controlled by man. Nevertheless drought can be of relevant importance in some areas.

Current land use/silviculture (including history and trends)

The poplar plantations are managed as conventional crops, using typical agronomic practices, chemicals etc. These plantations are usually located in flood plains and areas with high water table. In Italy they are concentrated in the north (Pianura Padana) and cover ca 70 000 ha (1% of the national forest area), providing 30–35% of the national timber needs. In the lowlands of northern Italy they typically occupy the natural distribution area of mixed oak-hornbeam forests. During the 1960s, on the central European flood plains, poplar plantations were planted on rich stream dependent sites, displacing species-rich natural riparian forests.

Robinia plantations

Characteristics, ecological conditions and main tree species:

Robinia is usually planted on sandy and loess areas on relatively dry sites, sometimes even on bare sand. It does not have special requirements concerning soil chemistry and nutrient supply. It grows best where the groundwater table is moderately high. *Robinia* is a light demanding species and has extremely powerful sprouting ability both from roots and stumps. It has symbiotic nitrogen fixating bacteria, so it effectively influences soil chemistry. For these reasons it seldom occurs in mixtures with other tree or shrub species (*Celtis occidentalis*, *Padus serotina*, *Ptelea trifoliata*, *Sambucus nigra*).

Main forest types according to country schemes

The question of naturalness is irrelevant concerning *Robinia* stands. Because of its aggressive behaviour it hardly mixes with any other species, only very few herbs can tolerate the conditions within a *Robinia* stand.

Since it is usually managed in a relatively short rotation, large trees and associated natural phenomena are usually lacking.

Robinia is at the moment extending its area aggressively. It is expected that some isolated forests, some of them quite large, will be substituted by pure *Robinia* forests already after two rotation periods. *Robinia* populations cannot be compared with plantations of poplar clones, which have lost the ability of natural regeneration, instead *Robinia* is characterised by an aggressive extension. Other less-competitive species will be outcompeted in these forests.

In Hungary *Robinia* was an important species used for the afforestations on the Great Hungarian Plains. Nowadays it covers almost 20% of forested land in the country, mostly in lowlands and in low, hilly areas. Its spreading ability causes serious conservation problems, and once established, it is extremely difficult to eradicate.

Further, several types are described in phytosociological terms, but they are not easily distinguished. It is, however, reasonable to distinguish types based on site characteristics, for instance water regimes. In Austria this type is not included in the national system, which is mainly based on natural or semi-natural forests.

Key factors of forest biodiversity

STRUCTURAL: Extremely simple structure, one canopy layer, optional shrub layer of *Sambucus nigra*, scarce herb layer of a few species, hence limited richness of associated fauna.

COMPOSITIONAL: At the driest sites elements of original steppe vegetation might survive.

NATURAL DISTURBANCES: Irrelevant.

Current land use/silviculture (including history and trends)

Short rotation (25–30 yr), different size of clearcuts, regeneration from sprouts, or where used after other forest types, cuttings are used for afforestation after complete soil preparation (removing trunks, deep ploughing etc.).

Eucalyptus plantation

Characteristics, ecological conditions and main tree species

These plantations cover a large area of several hundred thousand hectares in Portugal mainly in the last decades. In Italy, it occurs only in the south, in areas without high forest vegetation (Sicily, Campania and Lazio), in Spain it occupies ca 3% of total forested area. Further, in Greece it occurs only in the south as old stands. Main species are *Eucalyptus globulus* and *E. camaldulensis,* very productive species, water demanding and frost sensitive, preferring moist areas with mild winters.

Main forest types according to country schemes

No forest types can be applicable to these plantations, but it is usually included as a group in forest inventories.

Key factors of forest biodiversity

STRUCTURAL: The most important ones include age of stands, stand size and edge characteristics, understorey vegetation, as well as tree stand and structural diversity,

Eucalyptus plantation, photo: Peter Friis Møller.

which contributes to more or less heterogeneity of habitat types. Litter is also very important because of its special chemical composition and the fact that it is the basis for the chemical and physical environment of soil fauna. A more diverse structure can be achieved when elements of the original vegetation are included and maintained (shrubs, *Salix* spp.) and when rotation period is extended.

COMPOSITIONAL: In this type of plantations, the soil composition is of stressed importance because it determines the composition of the understorey vegetation and contributes to a particular environment where, for instance, soil fauna interact with the surroundings.

NATURAL DISTURBANCES: The main natural disturbances are related to insect pests (e.g. *Phoracanta* spp. *Xylophagous defoliator*) and soil erosion. Fire can occur but to a lesser extent than in other plantations (e.g. pine). Human influences: forestry and forest management are the most important key factors. The intensity of the rotation and of the removal of the understorey vegetation are important for the habitat types (and subsequently species) that can exist in such exotic plantations. Storm may be important as poplar stands or tree groups growing quickly may be higher than surrounding stands.

Current land use/silviculture (including history and trends)

Mainly for pulp production. The plantation area has not been significantly expanded since the last decade. Some of the areas are still intensively used, with rotations of ca 10 yr and artificial regeneration after four rotations. This forest type regenerates very well after fire as *Eucalyptus* competes successfully with pines and oaks when disturbances are very frequent. Also as it is not browsed by herbivores which adds to its competitive ability. In Madeira *Eucalyptus globulus* was introduced during the 19th century. Nowadays it has spread around the island mainly at private properties with no significant commercial value.

Chapter 5.

Principles for Assessing Forest Biodiversity in Europe

5.1 Biodiversity indicators and assessment

In this section we will briefly review the concept of biodiversity indicators, and try to put the methodology advocated by BEAR into a general context of current indicator concept development. The literature on biodiversity indicators was reviewed by Hansson (2000, Appendix 7). Furthermore, many ideas on this topic was presented in a recent Expert Meeting and Electronic Conference organised by Larsson and Esteban (2000), cf. Chapter 6. Operational guidelines on the assessment of key factors influencing European forest biodiversity are presented below in and Chapter 6.

An indicator may be a species, a structural component, a process or some other feature of the biological system, the occurrence of which insures the maintenance or restoration of the most important aspects of biodiversity for that system (Hansson 2000a) (cf. Appendices 6 and 7).

Indicators can be used to provide decision makers with useful information on the status of and trends in biodiversity, and to help determine if broad goals and targets for conservation are being reached (Reid et al. 1993, Barbati et al. 1999b, Del Favero 2000). When measured, they should repeatedly demonstrate ecological trends, and measure the state or quality of an area (Corona and Pignatti 1996, Ferris and Humphrey 1999). They can be quantitative or qualitative.

Ideally an indicator should be (Noss 1990):
• Relevant to ecological significant phenomena;
• Able to differentiate between natural cycles or trends and those induced by anthropogenic stress;
• Capable of providing a continuous assessment over a wide range of stress;
• Sufficiently sensitive to provide an early warning of changes;
• Distributed over a broad geographical area, or otherwise widely applicable;
• Easy and cost-effective to measure, collect, assay and/or calculate.

Because no single indicator is thought to possess all of the desirable properties required to make a full assessment of biodiversity, a set of complementary indicators is therefore required.

Biodiversity indicators may be chosen in two ways:
1) First, they may be PARAMETERS OF A PARTICULAR COMPONENT OF BIODIVERSITY. For example, a parameter of bird diversity (such as bird species richness) may be used as an indicator of bird diversity, a parameter of bird diversity may be used as an indicator of animal diversity or a parameter of tree diversity may be used as an indicator of overall plant and animal species diversity. The choice of the indicator depends on many factors but it always assumes that there is a correlation between the indicator and the component of biodiversity that it is being used to indicate. This assumption may be based on scientific studies or on a knowledge of natural history. In the latter case, the indicator may be said to be unverified but even in the former case the correlation is likely to be empirically derived and does not imply any true relationship between the indicator and the component of biodiversity being estimated. This type of indicator is, of course, always compositional (i.e. based on species or other components of the composition of the forest ecosystem) and a direct measure of biodiversity per se. Because of the way that these type of indicator are derived, they may be referred to as empirical indicators. They are usually related to measures of species diversity. Most texts on biodiversity indicators concentrate on this type of indicator. Indeed McGeoch and Chown (1998), in a recent review of biodiversity indicators, recommends that the term biodiversity indicator be used only to refer to species (taxon or taxa). McGeoch and Chown (1998) give the following definition: "A biodiversity indicator is a group of taxa…, or functional group, the diversity of which reflects some measure of the diversity… of other higher taxa in a habitat or set of habitats" (Gaston and Blackburn 1995, Gaston 1996). These indicators are most commonly measured by counting the number of plant or animal species present in a given area (species

richness), or, in addition, their relative abundance and evenness as part of a diversity index (species diversity). Single species indicators have been used, e.g. the Management Indicator Species enforced by US law for protecting original states in American forests (cf. Hansson 2000), but they have been criticised because of the poor relationship between the presence or absence of a particular species and biodiversity. As a consequence, indicator species "groups" are used more commonly – e.g. birds (Järvinen and Väisänen 1979) and tiger beetles (Pearson and Cassola 1992). However, several studies (Prendergast et al. 1993, Lawton et al. 1998) suggest that the correlation between the diversity of species groups taxa is low and that, therefore, reliance on a single indicator group would give a poor measure of overall biodiversity.

2) There is, however, a second way of choosing biodiversity indicators. It involves acknowledging that, in the forest context, biodiversity is dependent on the structure of stands and landscapes, the species in them and the management and disturbance regimes they experience. In other words, BIODIVERSITY INDICATORS MAY BE DERIVED FROM AN ANALYSIS OF THE KEY FACTORS AFFECTING BIODIVERSITY. Thus an indicator based on key factors of biodiversity may be either: structural, compositional (species etc.) or functional (management or disturbance). Although empirically-based compositional indicators may be usefully incorporated in a set of indicators, the derivation of indicators through key factors of forest biodiversity was the approach mainly adopted in the BEAR-project. A few examples from literature as regards indicators based on key factors of biodiversity:

STRUCTURAL INDICATORS: The use of structural indicators is based on the supposition that a more complex habitat will support a greater variety of species. Trees and stand structure have a major impact on other components of the forest ecosystem such as birds (MacArthur and MacArthur 1961, Moss 1978, James and Wamer 1982, Erdelen 1984), mosses and lichens (Rose 1992) and insects (Murdoch et al. 1972, Gunnarsson 1996). For this reason, stand structure, the dynamics and development of forest stands, can play a key role in the derivation of biodiversity indicators in undisturbed and managed forests and woodlands in Europe. Structural change for example may increase the forests' susceptibility to environmental disturbances and encourage the loss of native species through the establishment of invasive, non-native species. Canopy structure controls the quantity, quality, spatial and temporal distribution of light, local precipitation and air movements. Combined, these factors determine air humidity, temperature and to some extent soil moisture conditions (Jennings et al. 1999), which ultimately influence the microclimatic suitability for the regeneration of flora and fauna. Structural indicators might for example include cano-

py openness. Dead and dying wood is another important structural component (Ferris and Humphrey 1999). It is a key factor for forest biodiversity in forest ecosystems, particularly when a range of forms of deadwood are present (Harmon et al. 1986).

COMPOSITIONAL/SPECIES INDICATORS: Compositional indicators are usually empirical indicators as discussed above. However, a functional relation may exist as most species depend, at least to some extent, on the presence of other species. This is the basis for the concept of "keystone species" being "drivers" instead of "passengers" in the ecological systems (Walker 1992, Appendix 6). For example, tree diversity is a key factor affecting the diversity of many other groups and, can therefore, be used as an indicator of biodiversity. Hansson (2000) identifies three strategies to select compositional indicators: Firstly, species or physical structures that are strongly related to various stages or disturbance regimes of focal ecosystems could be used as indicators about the existence or lack of prerequisites for the majority of species. Secondly, nestedness of species matrices could be used to distinguish relatively rare species that usually occur when the majority of the species are present – however, still with the problem of poor correlation between taxa. A third way, advocated by Hansson (2000), is based on Lambeck's (1997) "focal species approach" of selecting a set of species sensitive to a) the size of area required; b) isolation and limited connectivity; c) crucial resources; d) natural or anthropogenic processes as recurrent fires, grazing etc. If a set of focal species having reasonably high demands in these respects is selected, and landscapes are managed to retain such species, then the vast majority of all species should also be given the prerequisites to exist. The BEAR project recommends this approach to selecting compositional indicators, cf. the further comments in connection to this in Chapter 5.2.

FUNCTIONAL INDICATORS: To assess the different scales of diversity within an area, an integrated approach is required that includes not only species abundance but also their functions, sizes, spatial distribution and other information (hence the use of functional groups, guilds, keystone species). Indicators of function or process (e.g. decomposition) are particularly valuable for use when assessing biodiversity in its fullest sense. For example, the state of decay of dead and dying wood, and humus form, provide an indication of decomposition and nutrient cycling. If decomposition is too slow, most of the nutrients returned to the forest floor are removed from active circulation for long periods of time, and both nutrient cycling and forest productivity are reduced (Kimmins 1997a).

In the recent Electronic Conference on "Cost-effective indicators to assess biological diversity in the framework of the Convention on Biological Diversity" organised by Larsson and Esteban (2000) a number of different measures of biodiversity were put forward. It was also interesting to note that the contributing experts created an extensive discussion on the implication of "assessing" biodiversity. Allan Watt, on behalf of the Conference organisation, in a concluding remark stated that "assessing" means more than just measure but should be interpreted "to estimate value of". More specifically Watt pointed at the definition by the SBSTTA (1997): "(Biodiversity) Assessment: analysis of the gap between the present state and a reference one". This is the ultimate objective of indicators for biodiversity assessment, however limited by the fact that at most preliminary reference levels may be identified with present knowledge (cf. the need for research discussed in Chapter 6.3).

The BEAR concept of "Biodiversity Evaluation Tools BETs" could consequently just as well be called "Biodiversity Assessment Tools" (but we suggest to hold on to BETs). The important thing to realise is that BETs are more than indicator schemes and preferred methodology to (more or less) "objectively" measure biodiversity – the BETs should aim at providing information for operational use in biodiversity planning and management.

5.2 Assessing the key factors of forest biodiversity

In Chapter 3.1 a preliminary list of key factors of forest biodiversity is presented, at national, landscape and stand scales. Although, having different relative importance, these factors are relevant to all European forest types. The concept of Forest Types for Biodiversity Assessment FTBA, i.e. forest types with specific characteristics with regard to key factors of biodiversity; was introduced in Chapter 3.2 and the European level FTBAs were described in Chapter 4. This approach has in principle been found acceptable in a mid-project consultation with the BEAR user-panel (Appendix 1).

In this section, the possibilities to assess each key factor are discussed, and optional lists of indicators are presented, adapted to biogeographic region (cf. Chapter 2.2) and methodological prerequisites (Chapter 6.2) when this is necessary. The reader interested in the background and relationship to other approaches of indicator development should consult the previous section (5.1) and Appendix 7. In Chapter 6.3 the actual state of knowledge and most urgent research needs are further discussed.

In the mid-project consultation with the BEAR user-panel several organisations expressed a concern that the BEAR project should not present a manual for the operational implementation of a large number of indicators. The project has even received the advice to stop at recommending biodiversity assessment based upon key factors and FTBAs. The solution to the dilemma caused by this

advice and the commitments in the project plan to present a manual "Biodiversity Evaluation Tools BETs" has in spite of the expressed concerns, also presented a set of rather extensive indicator schemes but also clearly stressed that the presented indicators are optional. The final choice is up to the user/stakeholders, also keeping in mind the fact that different user groups may have different preferences and practical prerequisites.

To make this key information readily available for the user ("the BETs manual"), a standard presentational style is used for indicators to assess structural, compositional and functional key factors; with respect to the national, landscape and stand level. Lists of optional indicators are given in tables; marking indicators that can be considered to have high ecological relevance and giving an overall judgement of data availability. Eight tables of indicators are presented, dealing with:

Structural indicators assessing in Table:
    7. structural key factors at the national scale;
    8. structural key factors at the landscape scale;
    9. structural key factors at the stand scale;

Compositional indicators assessing in Table:
    10. compositional key factors at the national scale;
    11. compositional key factors at the landscape scale;
    12. compositional key factors at the stand scale;

Functional indicators assessing in Table:
    13. functional key factors at all scales; natural influence;
    14. functional key factors at all scales; anthropogenic influence.

It should be noted that some key factors and indicators specified for the national, landscape or stand scale may also be useful at other scale(s). This probably applies particularly to national-level factors/indicators at the landscape scale. Furthermore key factors may in some cases be assessed by indicators grouped under a different heading (e.g. some species may be good indicators of certain functions like fire/lack of fire). In order to avoid repetition we have listed the indicators at the specific level and the key factor types we consider to be most appropriate.

Structural indicators

The present knowledge allows relatively extensive lists of potential indicators to assess structural key factors. Structural indicators are widely used in forest inventories and several in current use are more or less related to biodiversity. Therefore data availability is relatively good or can be collected at moderate costs. The methods to measure structural indicators range from national statistics, satellite observations, other remote sensing techniques to on-ground field observations. There is also a variation in the indicator methodologies currently in use,

depending on national prerequisites as regards forest and other biodiversity related inventories. The overview in Tables 7–9 below and comments is for the above reasons by no means complete. A general word of caution may also be in place as regards taking on board a long list of structural indicators when designing Biodiversity Evaluation Tools; over-lapping indicators and indicators with unclear relation to biodiversity and actual management issues should be avoided in favour of a more consolidated scheme that effectively can be further processed in the planning procedures.

Table 7. Assessment of structural key factors of forest biodiversity at national scale. The list of indicators is intended to give an overview of possibilities – cf. Chapter 6 for further recommendations for establishing indicator schemes.

KEY FACTOR	CONCEIVABLE INDICATORS **Bold text**: The ecological significance of the indicator is high. ** Good data availability and/or data can be collected with relatively small effort. * Data exist in some cases and/or data can be collected with medium effort.
Total area of forest with respect to forest types	**Forest land:** [1] **– Total area (ha) and in relation to total land area (%)**** **– Area (ha) with regard to forest types (CORINE habitat types, FTBAs)***
Area of productive forest with respect to tree species and age	Area (ha) of productive forest: [1] **– Total **** **– Forest land (ha) with regards to tree species composition (pure stand/species (groups), mixed stands/species (groups))**** **– Age classes***
Forest ownership	Forest area with regard to ownership: – Total area (ha) and % of owner category (state, church, local communities/municipalities, forest companies, other private incl. forest farmers) – Number/category and mean area/owner category
Total area of forest with respect to legal status/utilisation or protection	Forest area: **– Area (ha) of forest and number of reserves/mean size) in nature protection areas (IUCN categories)**** – Strict forest reserves [2] (number and area, ha)*
Total area of old growth forest and forest left for free development	**Old growth forest:** **– Total area (ha) and in relation to total forest area (%)*** – Protection status (IUCN)* **Area left for free development:** **– Total area (ha)*** **– Type: stand origin (natural regeneration, other), succession stage (e.g. mature forest, young stage, newly burnt)** – Protection status (IUCN)*
Total area of forest with respect to afforestation/deforestation	Total area (ha) of afforestation and deforestation/ desertification: **– Total since 1950 and per decade afterwards**** – Present yearly rate of afforestation/deforestation*

NOTES

[1] Note that the definition of forest varies. In some countries data are only available for forests of a certain productivity, while others follow e.g. the FAO definition, cf. definitions in Appendix 7. A good overview is provided by Parviainen et al. 2000. See also the comments in connection with Table 14 as regards the importance to monitor the total area subject to forestry.

[2] As defined by Pavviainen et al. 2000.

Table 8. Assessment of structural key factors of forest biodiversity at landscape scale. The list of indicators is intended to give an overview of possibilities – cf. Chapter 6 for further recommendations for establishing indicator schemes.

KEY FACTOR	CONCEIVABLE INDICATORS **Bold text**: The ecological significance of the indicator is high. ** Good data availability and/or data can be collected with relatively small effort. * Data exist in some cases and/or data can be collected with medium effort.
Habitat composition	**Presence of each type (FTBAs, CORINE) of habitats:** – **Area** [1] (ha, ha of type 1000 ha^{-1})
Lakes and rivers	**Lake/river presence:** – **Type**** (agreed classification scheme) and area/length – **Riparian zone type(s) and widths** **Regulation and pollution status:** – **Presence/absence of unconfined rivers/streams**** – Estimated volume organic debris per unit length
Spatial continuity and connectivity of important habitats	**Landscape connectivity:** [2] – **Corridor length and width** (agreed classification scheme and length/area)** – Proximity measures for different areas of particular habitats
Fragmentation	**Fragmentation measures:** – **Specific habitat (esp. forest) patchiness (patch size, absolute, % and rate of change)** related to base-line – Landscape graininess index related to base-line, (absolute and rate of change) – **Total edge/ecotone length** (m 1000 ha^{-1}), specify type **Forest roads*:** – Road type(s), abundance (km 1000 ha^{-1})
History of landscape use	**Qualitative and if possible quantitative description with respect to forest continuity:** [3] – Qualitative description of land-use favouring/disfavouring specific forest features (e.g. old trees, gaps, prescribed burning), hunting, grazing etc.*

NOTES

[1] The most readily-available landscape-scale forest-cover data will be on CORINE habitat types, but this uses very broad forest categories and have limited resolution in relation to small patches of habitat. Other data may be available for given landscapes in the form of forest inventories, and it should be possible to match these against standard FTBAs as defined for this project. Further categorisation of forest area (at least overall) into standardised age-classes (e.g. 1–10, 11–50, 50–100, 100–150, 150+ year-old trees) – which for some biota may provide effectively different habitats – may be possible using existing forest inventory data but may require further ground-truthing to refine remote-sensed data on forest cover.

[2] Connectivity implies not only corridors connecting areas of a habitat, but also the general ease of movement or (re)colonisation possibility between areas. To some extent, proximity is implicit in fragmentation measures.

[3] The area (and % area) of landscape that has been continuously forested for long periods, at least 100 yr, is likely to be of particular importance. Data (at least qualitative) on temporal continuity for standard FTBAs would also be valuable. Forest inventories are likely to provide suitable data in many cases, but will be of particular value for shorter time-scales and for commercially-exploited forest types.

Table 9. Assessment of structural key factors of forest biodiversity at stand scale. The table is largely composed of a selection of indicators which are standard forestry data, that may be collected using established methodology. Existing site classification procedures may include additional indicators. The selection of indicators must be cost-effective with regard to the specific purpose and biodiversity analysis to be performed – a warning should be given concerning extensive data collection for the description of stand structural indicators!

KEY FACTOR	CONCEIVABLE INDICATORS **Bold text**: The ecological significance of the indicator is high. ** Good data availability and/or data can be collected with relatively small effort. * Data exist in some cases and/or data can be collected with medium effort.
Tree species ("site original", "not site original" and non-native)	**Species:** **– Volume/biomass (total m³, % per species)**** **– Basal area, m², % (or numbers)**** **Presence and spatial distribution of different tree size and age classes:** **– Numbers**, density, clustering measures**
Stand size	ha**
Stand edge/shape	**Stand shape:** **– Edge to area ratio**** **Ecotone:** **– Type**** **– Surrounding habitat****
Forest history	**Stand continuity:** **– Indicator species*** **– Historical maps*** **– Area of old growth/ancient forest*** **– Area of recent forest*** **– Area of plantations***
Habitat type(s)	**Area (%) of different habitat:** **– Agreed classification schemes**** **– National forest types****
Tree stand structural complexity (horizontal and vertical)	**Horizontal structure:** **– Gap occurrence**** **– Tree clustering*** **Vertical structure:** **– Layering (single, multi, understorey)**** **– Canopy layering (even, undulating)*** **– Shrub-layer (% area)*** **– Natural regeneration (seedling density)***
Dead wood	**Presence, nature and spatial distribution of standing and lying dead wood:** **– Type (snag, lying position), species, decay class and amount (volume, diameter distribution, basal area)****
Litter	**Humus:** **– Type or quality of humus form, amount (cm)*** **Flammable litter:** **– Amount (cm)***

Compositional indicators

Direct assessment of biodiversity composition ("species") may for obvious reasons be attractive. Several approaches exist, see Chapter 5.1 for a discussion of alternatives and BEAR preferences on a principal level. Data availability and/or cost for collecting data on compositional indicators is crucial. The BEAR project recommends in the short-middle term time perspective compositional indicators only as a complement to indirect assessment through structural and functional indicators, see also Chapter 6. This recommendation is mainly based upon the fact that there is very little existing knowledge to validate compositional indicators as regards a wider range of biodiversity, the interpretation of absence of a certain indicator species, time aspects etc, cf. Chapter 6.3.

A particular case are introduced, non-native species. This phenomena is from many points of view a serious threat to biodiversity, in some parts of the world and/or locally of the same magnitude as habitat change caused by change in land-use. The plantation of forest trees outside, but not far from, it original species distribution area or on sites outside the normal ecological range may be a similar problem. The issue of non-native and not site-original species has not been possible to deal with in depth in the BEAR project, but indicators are outlined in the tables below.

Table 10. Assessment of compositional key factors of forest biodiversity at national scale. The list of indicators is intended to give an overview of possibilities – cf. Chapter 6 for further recommendations for establishing indicator schemes.

KEY FACTOR	CONCEIVABLE INDICATORS
	Bold text: The ecological significance of the indicator is high.
	** Good data availability and/or data can be collected with relatively small effort.
	* Data exist in some cases and/or data can be collected with medium effort.
Native species	**Red-listed species**: [1]
	– Status of selected taxa, status and trends (IUCN categories), national level or biogeographic region **
	– Status, population numbers and trends selected target species at national scale or biogeographic region, for each FTBA*
	Other forest species: [2]
	– Trend estimates for selected forest species, for each FTBA
	– Bird monitoring data (trend indication)*
	– Game statistics (trend indication)*
	– Lichen monitoring (trend indication)*
Non-native or not "site original" tree species	**Forest trees**:
	– Area and yearly rate of introduced tree species (ha/species)**
	– Area and yearly rate of plantation of not site original species (ha/species)*
	Other species:
	– Area covered by selected introduced understorey species
	– **Number of other non-native forest species in selected taxa and forest types****

NOTES

[1] Most countries are able to present the state of red-listed species of a number of taxa, according to an agreed nomenclature. These, data must, for use in this context, separate the "forest species". This will create a number of border-line cases that need to be dealt with in a harmonised way. The BEAR project welcomes in particular red-lists showing status of selected forest species for each Forest Type for Biodiversity Assessment. For the European perspective these lists should be based upon threat status in the Biogeographic Regions (developed in accordance with the Habitats Directive). This type of information is particularly valuable for trend analysis (which is already possible in some cases). A major problem with red-lists is that they are based upon "expert opinions" based upon heterogeneous knowledge, that can be expected to change with time.

[2] The population development of dominating forest species may be at least as important to follow as the status of red-listed species. Unfortunately there are hardly European-scale data available on other forest species than the red-listed ones. Several forest game species could be of interest in this respect, but European-scale efforts to harmonise hunting statistics on forest game have so far failed. A possibility is to introduce qualitative assessment of population status and development of selected more common forest species as a complement to the red-list – a high priority development project. On a national scale some countries have good, quantitative, data showing temporal development of certain more common forest species; e.g. the precise game statistics of Denmark, the national system of "Game triangles" in Finland, bird monitoring schemes in several countries etc. National schemes to monitor forest biodiversity or aiming at identifying forests with biodiversity values, are based upon indicator lichens, fungi, vascular plants, moths and other insects collected by light traps etc. As data collection may be quite extensive (e.g. in Switzerland, Belgium, Sweden, Finland and U.K.) this may be used to demonstrate biodiversity development on national and landscape scale.

Table 11. Assessment of compositional key factors of forest biodiversity at landscape scale. On a landscape scale compositional indicators, i.e. species with specific landscape-scale requirements, have a great potential to monitor and verify the success of planning measures. Biodiversity Evaluation Tools for landscape-scale planning of biodiversity are recommended to include "focal species" (see chapter 5.1) selected in dialogue with the planners/managers (see chapter 6.2) based upon advice from biological expertise, taking data availability and costs for collection into account. In the table some options are listed.

KEY FACTOR	CONCEIVABLE INDICATORS **Bold text**: The ecological significance of the indicator is high. * Data exist in some cases and/or data can be collected with medium effort.
Species with specific landscape-scale requirements	**Top predators, selected birds (woodpeckers, game birds):** [1] – **Game statistics** (population indices, trends)* – **Bird monitoring data** (population size/indices/ trends)* – **Other available data** (qualitative or quantitative estimates of population size and trends)
Non-native or not "site original" tree species	**For each non-native species:** – **Species, area, increase/decrease area/year** **For each not site "original" species:** – **Description of "problem" e.g. plantation outside original distribution area in country, plantation on special site e.g. peatland** – **Species, area, increase/decrease area/year**

NOTE

[1] Game statistics on the landscape scale should in many areas not be difficult to collect given a good co-operation with hunters organisations, the same apply to bird and other biological data (and relevant organisations).

Table 12. Assessment of compositional key factors of forest biodiversity at stand scale.

KEY FACTOR	CONCEIVABLE INDICATORS **Bold text**: The ecological significance of the indicator is high. * Data exist in some cases and/or data can be collected with medium effort.
Species with specific stand type and scale requirements	**Species presence/absence*** [1]
Biological soil condition	**Humus type**

NOTE

[1] A number of species/taxa are applied as indicators in stand-scale assessment of forest biodiversity – vascular plants, bryophytes, lichens, fungi, "soil fauna", insects, amphibians (in US), birds – and species inventories are a "must" in any Biodiversity Evaluation Tool on the stand scale. The selection of species, for different Forest Types for Biodiversity Assessment, could be made in dialogue with national/regional biological expertise, following "the focal species approach" presented in Chapter 5.1.

Functional indicators

The BEAR project strongly recommends including functional indicators in any Biodiversity Evaluation Tool. Data availability on a national scale may not be so good, but semi-quantitative measures and even expert opinions would be most valuable for European-scale assessments. Biodiversity Evaluation Tools on landscape level should give high priority to collect information on functional indicators while this type of stand-scale information should be readily available in local forest plans or other local knowledge.

Table 13. Assessment of functional key factors of forest biodiversity at all scales: natural influence. The list of indicators is intended to give an overview of possibilities – cf. Chapter 6 for further recommendations for establishing indicator schemes. Most data on natural disturbances; fire, wind and snow damage as well as insect outbreaks are at the present to be considered mainly as "event statistics", ecologically relevant only to show the very first successional stage (as the "damaged" forests as a rule are quickly taken care of). Areas left for free development after natural disturbance are extremely valuable to record, at all scales.

KEY FACTOR	CONCEIVABLE INDICATORS **Bold text**: The ecological significance of the indicator is high. * Data exist in some cases and/or data can be collected with medium effort.
Fire [1]	**Area of forest burnt (over specified time period), preferably for each FTBA*** Percentage area of protective forest burnt* Yearly area of regeneration (ha, %), (specified as planted and natural regeneration)
Wind and snow	**Area of forest affected e.g. windblows (over specified time period)*** Percentage area of protective forest affected Yearly area of regeneration (ha, %), (specified as planted and natural regeneration)
Biological disturbance	**Area affected by specified disturbance** Changes in humus form over specified period*

NOTE

[1] The occurrence of fire, being a most controversial, albeit important, key factors of forest biodiversity, should however, be monitored at all scales. Due to effective fire control northern forest types "lack" successional stages created by forest fire. On the opposite fire is generally considered a problem in the Mediterranean due to a high number of anthropogenically induced fires. In both areas "reference states" urgently needs to be identified. Restoration measures, e.g. the use of prescribed and nature conservation fires in northern forests as well as fire prevention in the Mediterranean, should also be monitored in particular at landscape scale.

Table 14. Assessment of functional key factors of forest biodiversity at all scales; anthropogenic influence. The list of indicators is intended to give an overview of possibilities – cf. Chapter 6 for further recommendations for establishing indicator schemes.

KEY FACTOR	CONCEIVABLE INDICATORS **Bold text**: The ecological significance of the indicator is high. ** Good data availability and/or data can be collected with relatively small effort. * Data exist in some cases and/or data can be collected with medium effort.

Forestry; "normal" silviculture	**Area and % of different silvicultural systems, specifying, as appropriate:**
	    – Yearly area clearcut (ha, %)**
	    – Yearly regeneration area (ha, %), specify planted and natural regeneration, prescribed burning, draining**
	    – Yearly area shelterwood system**
	    – Yearly area coppice system**

Forestry, "specific" silviculture	**Area and % of specific silviculture:**
	    – Protection forest (specify type, e.g. wind/erosion, climate, water protection) **
	    – Nature conservation forestry and/or forest with prolonged rotation period**
	    – Urban/recreation forestry*

Agriculture and grazing	**Forest areas under different agricultural systems (ha);**
	    – Agroforestry (specify e.g. cork oak, olive, ha)**
	    – Other land with significant agricultural/grazing utilisation combined with trees/forest (specify, ha, number and mean density 100 ha^{-1} of grazing animals)*
	Areas with significant remaining impacts from former agriculture/grazing (specify) *

Other land-use	**Specify (e.g. military, hunting area, park-land)****
	Social influences:
	    – Visitor pressure (numbers)
	    – Establishments
	    – Amount of garbage left in forests (volume, type)

Pollution	**Forest areas (ha) with impact of different types of pollution (in categories in relation to critical loads or levels) *** **Water:** – Sodium dominance index

GENERAL COMMENT: Changes in land-use are considered to be the main recent and current threat to European forest biodiversity, see Chapter 3.2. Forestry, which is in focus in this report, is performed on most forested land. The utilised area may vary due to economic and other societal prerequisites, nature protection and other restrictions. The development of the area subject to active forestry is thus of major interest in a biodiversity monitoring scheme. Mainly because of common inventory techniques, the indicators demonstrating changes in areas subject to forestry are dealt with as structural indicators while we here point at a need to further specify the type of forestry and silvicultural methods. This is of course also of interest to understand the prerequisites for the development of forest biodiversity, although the functional relationships may not be known in all cases. During recent decades, forestry in most countries have introduced alternative silvicultural methods and strategies to include biodiversity in the sustainability concept. Monitoring of areas treated with "specific" silvicultural schemes should be introduced according to harmonised concepts.

Agricultural changes of direct relevance to forest biodiversity occur at present in Europe, and should be monitored at all scales. The process of active reforestation of abandoned agricultural land has, like the forestry development, been dealt with as a structural indicator. The Mediterranean "agroforestry" types (oak, olive) containing trees of great biodiversity interest are diminishing/deteriorating. In a monitoring of forest biodiversity this can not be neglected, although the management is not the responsibility of the forester/forest policy maker. The Mediterranean landscape has since historical times been heavily (over)grazed and many habitats have been shaped under this pressure. Extensive grazing is still a great, but diminishing, phenomenon that is of great concern to south European forestry, but also to forestry in some regions (e.g. domestic grazing in Norway, winter grazing of reindeer in Fennoscandia). Monitoring of the development of extensive grazing is of high policy relevance on a European, national and landscape scale. The evaluation of the impact of (heavy) grazing is not easy from a biodiversity point of view, as long as agreed references do not exist.

Finally, being widely neglected in this context, some more detailed comments as regards indicators of pollution seem justified:

KEY FACTOR: Pollution.

INDICATOR: Forest, catchment and freshwater areas (ha, % area, or stream/length, % length) with impact of different types of pollution (in categories in relation to critical loads or levels, including forest-mediated acidity influences in sensitive catchments).

INDICATOR: Sodium dominance index. This is an indicator of acid sensitivity in rivers, particularly where contributions of salt from sea water dominate the base cation input.

These indicators are intended to reflect possible adverse pollution influences on the forests themselves, that might negatively impact biodiversity, and also possible forest-mediated negative impacts on freshwater chemistry, that might negatively impact aquatic biota within forested catchments. The latter issue relates particularly to catchments where there is or has been significant afforestation and where the underlying geology provides limited buffering capacity against conifer-mediated acidification.

Chapter 6.

Conclusions

6.1 A strategy for assessment of forest biodiversity on a European scale

The general background for European and national-scale assessment of forest biodiversity is briefly described in Chapter 1.2. The political, legal, economic and other factors which need to be taken into account when suggesting improvements in the systems used e.g. in reporting biodiversity status to the Convention on Biological Diversity or within the framework of this or other activities co-ordinated by the European Environment Agency or Commission Environment Directorate are rather complex.

We present some general advice on the use of BEAR results, proposing an ordering of priority actions, for application in European and national scale assessment and monitoring of forest biodiversity status

Recommendation Number 1: Introduce the key factor approach for European and national scale monitoring of forest biodiversity

The key factor concept as introduced in Chapter 3 is crucial of monitoring the prerequisites for forest biodiversity (structural and functional key factors) and main elements of biodiversity (compositional key factors). In practice, a summary of the state and trends in a set of key factors should be the most efficient measure of the state and trends in forest biodiversity. The following specific actions should be followed:
- A priority list of key factors should be agreed on, based upon the Table 4 in Chapter 3 and the Tables 7, 10, 13 and 14 in Chapter 5;
- Indicators based on these key factors must not necessarily be standardised. There are two main reasons for this:
  1. Data availability in different countries may vary. It is better, at least at an introductory phase, to allow semi-qualitative assessments and a range of indicators and methods than omitting an important key factor.

2. In general, a further development of indicator schemes and methodology is desirable.

Recommendation Number 2: Make a further division into different Forest Types for Biodiversity Assessment (FTBAs) in the reporting of each key factor

A great improvement in the reporting of state and trends of forest biodiversity can be brought about through the assessment of key factors with respect to the Forest Types for Biodiversity Assessment (FTBAs). The BEAR project has introduced these types to correspond to the key factor approach (cf. Chapters 3 and 4) and strongly recommends the adoption of this system.

The list of Forest Types for Biodiversity Assessment (FTBAs) is tentative, but the following should be taken into account when implementing this scheme:
- The POTENTIAL distribution of the FTBAs, can be illustrated by the Maps of Potential Natural Vegetation, as shown in Chapter 4. However, a quantitative assessment of the actual area of each FTBA is a prerequisite for presenting quantitative estimates of indicators relating to key factors. Developing methods for mapping, on a national scale, the actual area of each FTBA/country is a high priority task, to be followed by the collection of national FTBA data.
- Compatibility with other systems, e.g. CORINE, should be ensured, as long as the basic principles of deliniating the FTBAs are not violated;
- The FTBAs should not be further clustered without a full awareness of the increased heterogeneity with respect to the key factors.

In the long term, a more nature-based approach to European-scale biodiversity management would be to abandon the countries as units in favour of biogeographic regions, cf. Chapter 2.2.

Recommendation Number 3: Standardise indicators, methodology and protocols

Taking a long-term perspective, it is necessary for a standardised system of indicators of forest biodiversity to be implemented in a global assessment of state and trends in biological diversity. Europe may take the lead in this work, with the aid of the strong co-ordination by the European Commission. We advocate giving priority to the first two recommendations above, but propose the development of standardised indicators along two lines:

• Expand the National Forest Inventories (NFIs) to encompass indicators of forest biodiversity. This has several major advantages:

1. The NFIs are well established activities, designed and tested in operational use, and include quality assurance procedures with regard to inventory methodology, data handling and presentation. An extensive methodological literature exists.
2. The NFIs can often be performed according to a multi-level approach, including both remote sensing and field/ground measurements.
3. Indicators of forest biodiversity can, as a rule, be included at known costs (in some cases after some relatively moderate methodological tests).
4. The NFIs are ideally designed to reflect CHANGES in the state of selected indicators. The design of an NFI is often made to reflect changes at 5–10 yr periods (having a moderate, annual commitment to measurement) which, from an ecological point of view, is optimal (since it is necessary to eliminate annual variation from the analyses).

• Introduce satellite-based techniques for monitoring forest biodiversity as a support to the NFIs. Satellite mapping has the advantage of generating data which can be handled and analysed using GIS. This is probably the only feasible strategy for large-scale monitoring of forest biodiversity. We are close to operational use of satellite monitoring of forest biodiversity. Development projects have been inaugurated and these should be followed, taking into account:

1. The problem of satellite imagery techniques being "technology driven". This could be balanced by following the BEAR recommendations with regard to key factors and FTBAs.
2. A combination of monitoring techniques/inventory strategies should be used, to optimise cost-effectiveness.

6.2 Assessment of forest biodiversity on the operational scale

There are several threats to the biodiversity in forests: air(borne) pollution, water and soil pollution, climate change, other land-use (fragmentation/habitat loss), introduced species, frequent fires, recreation pressure incl. tourism etc. We will here focus on the forestry sector – taking nature conservation into account – and management of biodiversity at the operational scale; i.e. the silvicultural and logging activities of the forest managers.

European forestry has a well-developed silvicultural tradition and sustainability awareness. During the last decades the sustainability concept, in forestry traditionally with focus on the wood supply, has been enlarged to cover a number of societal needs. During the last decades, a number of biodiversity considerations are implemented in most silvicultural and management programmes. Although there is still a certain lack of scientific verification there is no doubt that forestry in many parts of Europe nowadays is better adapted to the site conditions and performed not only with a short rotation objective of maximising the wood production and consequently overexploitation.

Planning for biodiversity on the operational scale may occur at several spatial scales. This reflects not only the biodiversity components (genes, populations, species and ecosystems) but the fact that all these components are affected by processes at different scales. From an operational forestry point of view it is adequate to discuss three scales: Single tree, stand and landscape. One may distinguish three categories of "forest managers" that show different major "responsibilities":

Actor/ Biodiversity level	Small-scale forestry	Large-scale forestry	State/ regional
Single tree	••	•	
Stand	•	••	
Landscape		•	••

Single tree management

Without doubt, the small-scale forester, if he is the actual forest manager, has the best possibility to preserve single trees (particularly overaged, dying and dead trees) that are valuable from biodiversity point of view. He would certainly benefit from detailed advice on identifying valuable trees and how to manage them. Large-scale foresters, as a rule, also preserves valuable trees, but this has for practical reasons to be performed in a more stereotype way.

For single tree management the Biodiversity Evaluation Tools and indicators presented at stand scale in Chapter 5 are rather crude. They can serve as an important starting point but more specific measures need to be developed.

Dead wood may also be an illustrative example. The most valuable dying tree from the point of view of e.g. certain insects is the one that dies slowly, part by part, offering optimal substrate for wood-living insects. On the other hand, dead wood has many other biodiversity functions that could be taken into account.

Stand level management

Stand level management (in a broad sense) is basic knowledge to any forester. The small-scale forester has a more limited range of options for preserving stand-scale biodiversity, e.g. through setting aside valuable stands; has lower professional potential to adopt suggested indicators, and does not have the same tradition for undertaking detailed forest planning. However, this is a generalisation, since there are 12 million European small-scale forest owners, who will vary in their experience and may have the possibility to use external expertise.

The Biodiversity Evaluation Tools and indicators presented in Chapter 5 provide rather specific options for assessing forest biodiversity at the stand scale as a basis for developing and evaluating silvicultural methods. It is necessary to underline that the distinction between stand-scale management and landscape planning may be less clear in reality in view of the different scales relevant for different forest species. Taking a long-term perspective, only extremely large stands can successfully be managed alone (although under these circumstances they can be considered as landscapes). Recently, due to various subsidies provided by the states and EU, the silvicultural system of single tree selection (stand scale) has been introduced and applied by some small-scale forest owners.

Landscape planning

Recognition of the importance of biodiversity conservation at large spatial scales has led to an increased interest in landscape planning, i.e. optimising stand management on a landscape scale. Up to now small-scale forestry has not had an established tradition for undertaking collaborative landscape planning. Several large land-owners, including states, who can control forestry on the landscape scale, have introduced ecological and biodiversity landscape planning into their forestry programmes. It is important to recognise the responsibility of the state in preserving entire landscapes (as National Parks or large Forest Reserves) in more or less undisturbed conditions.

The Biodiversity Evaluation Tools and indicators presented in Chapter 5 are well suited for biodiversity planning at the landscape scale. Furthermore some indicators may be used to monitor key factors of forest biodiversity (a combination of structural, compositional and, in particular functional/natural disturbance) in National Parks or Forest Reserves. Core areas of National Parks or total areas of strict Forest Reserves could act as baseline reference points, at least at the stand scale (only in Fennoscandia may be found clusters of Forest Reserves and/or National Parks that cover sufficient large area to be considered as a landscape).

6.3 The need for further development and research

This report has avoided emphasising the uncertainties in connection with biodiversity indicators. However, all species are unique, and occupying specific niches in specific areas of an ecosystem. Consequently, the biodiversity indicators concept has a questionable scientific basis. Nevertheless, the introduction of biodiversity indicators is generally advocated, e.g. in the context of the Convention on Biological Diversity CBD.

The BEAR consortium actively participated in an Expert Meeting and Electronic Conference in 1999 on "Cost-effective indicators to assess biological diversity in the framework of the Convention on Biological Diversity (CBD)". A general consensus was reached as regards the development of operational indicators of biodiversity and concluded that "at present there is a strong development in this field and it is premature to establish standard monitoring schemes based on "core indicators" (Larsson and Esteban 2000). This is very much the philosophy of this report – a work strategy based on key factors and Forest Types for Biodiversity Assessment is recommended, but in the short perspective a variety of approaches as regards selecting indicators and measurement techniques is acceptable.

The needs for further research associated with European forest biodiversity was highlighted in an Electronic Conference in 1998 followed by ca 500 participants (Esteban et al. 1999). The BEAR consortium organised the session on forests, and the need to develop indicators was recognised in several sessions. Five priority research themes on forest biodiversity were identified (cf. Appendix 8):
1. Development of silvicultural methods;
2. Afforestation in agricultural or industrialised landscapes;
3. Management of forest fires;
4. Management of genetic resources in forests;
5. Management of protected areas.

Finally, five priority research areas are presented for the further development of Biodiversity Evaluation Tools for European forests, and their introduction into operational use:

1. VALIDATION OF INDICATORS OF FOREST BIODIVERSITY: Studies to validate indicators are urgently needed. This requires investigation of their predictive value with regards to components of biodiversity, ideally in a process allowing identification of optimal indicators.

2. ESTABLISH REFERENCE VALUES AND CRITICAL THRESHOLDS FOR FOREST BIODIVERSITY: This represents a logical further step once indicators have been selected. One reference situation is the pre-agricultural landscape (natural state). Fortunately, this is relevant for many European forests and a reconstruction of the state of biodiversity, as reflected by indicators, would therefore be of practical interest. When setting biodiversity targets (expressed as indicator values) critical thresholds in deviation from reference values are of great interest. Methods for identifying thresholds exist in landscape ecology and conservation biology. The concept of potential natural vegetation (Tüxen 1956) is valuable in this respect.

3. BIODIVERSITY EVALUATION TOOLS FOR ASSESSING GENETIC DIVERSITY OF FOREST TREES: The BEAR project has not considered genetic aspects, but research and data compilation in this fields is certainly needed, see Appendix 8.

4. DEVELOPMENT OF INDICATORS FOR NATIONAL MONITORING OF FOREST BIODIVERSITY: The National Forest Inventories should be expanded to encompass biodiversity measurements, cf. Chapter 6.1

5. STRENGTHENING LANDSCAPE-LEVEL PLANNING FOR FOREST BIODIVERSITY: There is a particular need to develop participatory planning procedures for landscape-level planning for the management of biodiversity in small-scale forestry.

References

Aichinger, E. 1967. Pflanzen als forstliche Standortszeiger. – Agrarverlag, Wien.

Alves, J. et al. 1998. Habitats naturais e seminaturais de Portugal Continental. Tipos de habitats mais significativos e agrupamentos vegetais caracteristicos. – ICN, Instituto da Conservação da Natureza, Lisboa.

Andrén, H. 1997. Habitat fragmentation and changes in biodiveristy. – Ecol. Bull. 46: 171–181.

Angelstam, P. 1992. Conservation of communities: the importance of edges, surroundings and landscape mosaic structure. – In: Hansson, L. (ed.), Ecological principles of nature conservation. Elsevier, pp. 9–70.

Angelstam, P. 1997. Landscape analysis as a tool for the scientific management of biodiversity. Boreal ecosystems and landscapes – structures, functions and conservation biodiversity. – Ecol. Bull. 46: 140–170.

Angelstam, P. K. 1998a. Towards a logic for assessing biodiversity in boreal forest. – In: Bachmann, P., Köhl, M. and Päivinen, R. (eds), Assessment and biodiversity for improved forest planning. European Forest Institute (EFI), Proceedings no. 18, pp. 301–313.

Angelstam, P. K. 1998b. Maintaining and restoring biodiversity in European boreal forests by developing natural disturbance regimes. – J. Veg. Sci. 9: 593–602.

Angemeier, P. L. and Karr, J. R. 1994. Biological integrity versus biological diversity as policy directives. – BioScience 44: 690–697.

Anon. 1984. Vegetationstyper i Norden. – Berlings, Arlöv.

Anon. 1989. Libro rojo de los bosques españoles. – ADENA/WWF España, Madrid.

Anon. 1990. Segundo Inventario Forestal Nacional. Explicaciones y métodos (1986–1995). – Instituto Nacional para la Conservación de la Naturaleza (ICONA), Ministerio de Agricultura, Pesca y Alimentación, Madrid.

Anon. 1996. Arbeitskreis Standortskartierung in der Arbeitsgemeinschaft Forsteinrichtung (Hrsg.): Forstliche Standortsaufnahme. Begriffe, Definitionen, Einteilungen, Kennzeichnungen, Erläuterungen. Kapitel B4: Die Vegetation als Ausdruck des Standorts. (Forest site assessment. Terms, definitions, classification, characteristics, comments. Chapter B4: The vegetation as expression of the site). Revised by Jahn, G. and Hübner, W. 5th ed. – Eching bei München, IHW-Verlag, pp. 193–241.

Anon. 1998. The Portuguese forest by numbers. – Ministério de Agricultura, do Desenvolvimento Rural e das Pescas.

Anon. 1999. Development of a standard approach to habitat classification in Ireland. – Unpubl. draft report, The Heritage Council, Kilkenny.

Arnborg, T. 1964. Det nordsvenska skogstypsschemat. – Sveriges skogsvårdsförbund, Stockholm, 6th ed., in Swedish.

Athanasiadis, N. 1985. Forest phytosociology. – Giachoudi-Giapouli Publ., Thessaloniki.

Atlegrim, O. and Sjöberg, K. 1996. Response of bilberry (*Vaccinium myrtillus*) to clear-cutting and single-tree selection harvests in uneven-aged boreal *Picea abies* forests. – For. Ecol. Manage. 86: 39–50.

Avery, M. and Leslie, R. 1990. Birds and forestry. – T. and A. D. Poyser.

Barbati, A. et al. 1999. Developing biodiversity assessment on a stand forest type management level in north-eastern Italy. – BEAR Technical Report 5, Swedish Environmental Protection Agency, Stockholm, Sweden.

Bartha, D. et al. 1995. Hazai erdőtársulásaink. (Forest communities of Hungary). – Tilia 1: 8–85.

Bass, S. 1998. Forest certification – the debate about standards. – Rural Development Forestry Network Paper 23b: 1– 21.

Berg, A. et al. 1994. Threatened plant, animal and fungus species in Swedish forests-distribution and habitat associations. – Conserv. Biol. 8: 718–731.

Bergeron, Y., Leduc, A. and Li, T.-X. 1997. Explaining the distribution of *Pinus* spp. in a Canadian boreal insular landscape. – J. Veg. Sci. 8: 37–44.

Bergeron, Y. et al. (eds) 1998. Key issues in disturbance dynamics in boreal forests. – J. Veg. Sci. 9: 463–610.

Björse, G. and Bradshaw, R. H. W. 1998. 2000 years of forest dynamics in southern Sweden: suggestions for forest management. – For. Ecol. Manage. 104: 15–26.

Blaschke, H. and Baumler, W. 1989. Mycophagy and spore dispersal by small mammals in Bavarian forests. – For. Ecol. Manage. 26: 237–245.

Blondel, J. and Aronson, J. 1999. Biology and wildlife of the Mediterranean region. – Oxford Univ. Press.

Bohn, U. 1994. International project for the construction of a map of the natural vegetation of Europe at a scale of 1:2.5 million – its concept, problems of harmonization and application for nature protection. – Colloques Phytosociologiques XXIII, Large area vegetation surveys, Bailleul.

Bohn, U. 1995. Structure and content of the vegetation map of Europe (scale 1:2.5 m) with reference to its relevance to the project entitled "European Vegetation Survey". – Annali di Botanica LIII: 143– 149.

Borhidi, A. (ed.) 1996. Critical revision of the Hungarian plant communities. – Janus Pannonius Univ., Pécs.

Boyle, T. J. B. et al. 1998. Criteria and indicators for assessing the sustainability of forest management: a practical approach to assessment of biodiversity. – CIFOR, Indonesia.

Bradshaw, R. H. W. 1998. The long-term control of forest biodiversity. – SkogForsk 1: 31–32.

Bradshaw, R. H. W. and Holmqvist, B. H. 1999. Danish forest development during the last 3000 years reconstructed from regional pollen data. – Ecography 22: 53–62.

Bradshaw, R. H. W. and Mitchell, F. J. G. 1999. The palaeoecological approach to reconstructing former grazing-vegetation interactions. – For. Ecol. Manage. 120: 3–12.

Bradshaw, R. H. W., Tolonen, K. and Tolonen, M. 1997. Holocene records of fire from the boreal and temperate zones of Europe. – In: Clark, J. S. et al. (eds), Sediment records of biomass burning and global change. Springer, pp. 347–365.

Braun-Blanquet, J. 1952. Les groupements vegetaux de la France Mediterraneenne. – Macabet Freres. Vaison la romaine.

Braun-Blanquet, J. 1964. Pflanzensoziologie, 3. Aufl. – Fischer.

Braun-Blanquet, J. and Tüxen, R. 1952. Irische Pflanzengesellschaften. – Veroff. Geobot. Inst. Zurich 25: 224–415.

Bunnell, F. L. and McLeod, C. A. 1998. Forestry and biological diversity: elements of the problem. – In: Bunnel, F. L. and Johnson, F. (eds), Policy and practices for biodiversity in managed forests: the living dance. UBC Press, Vancouver, pp. 3–18.

Burschel, P. 1992. Deadwood and forestry. – Allgemeine Forst Z. 47: 1143–1146.

Bücking, W. 1998. Comparison between managed and unmanaged forests. – SkogForsk 1: 14–15.

Cajander, A. K. 1949. Über Waldtypen. – Fennia 28: 1–176.

Calow, P. (ed.) 1998. The encyclopedia of ecology and environmental management. – Blackwell.

Ceballos, L. et al. 1966. Mapa Forestal de España. Escala 1:400 000. – Ministerio de Agricultura, Madrid.

Christensen, N. L. and Emborg, J. 1996. Biodiversity in natural versus managed forest in Denmark. – For. Ecol. Manage. 85: 47–51.

Corona, P. and Pignatti, G. 1996. Assessing and comparing forest plantations proximity to natural conditions. – J. Sustainable For. 4: 37–46.

Costa, J. C. et al. 1998. Biogeografia de Portugal Continental. – Quercetea 0: 1–56.

Cross, J. R. 1998. An outline and map of the potential natural vegetation of Ireland. – Appl. Veg. Sci. 1: 241–252.

Currie, F. A. and Bamford, R. 1982 The value of birdlife of retaining small conifer stands beyond normal felling age within forests. – Quart. J. For. 76: 153–160.

Dafis, S. 1976. Classification of forest vegetation of Greece. – Min. of Agricult., Gen. Direct. of For., Publ. no. 36, Athens.

D'Antonio, C. M., Dudley, T. L. and Mack, M. 1999. Disturbance and biological invasions: direct effects and feedbacks. – In: Walker, L. R. (ed.), Ecosystems of disturbed ground. Ecosystems of the World 16. Elsevier, pp. 413–452.

Davis, M. B. and Botkin, D. B. 1985. Sensitivity of cool-temperate forests and their fossil pollen record to rapid temperature change. – Quat. Res. 23: 327–340.

Del Favero, R. (ed.) 2000. Biodiversità e indicatori nei tipi forestali del Veneto. – Commissione Europea, Accademia Italiana di Scienze Forestali, Regione del Veneto, Mestre, Italy.

Dempster, J. P. 1991. Fragmentation, isolation and mobility of insect populations. – In: Collings, N. M. and Thomas, J. A. (eds), The conservation of insects and their habitats. 15th Symp. Roy. Entomol. Soc. Lond. Academic Press, pp. 143–153.

Denslow, J. S. and Spies, T. 1990. Canopy gaps in forest ecosystems: an introduction. – Can. J. For. Res. 20: 619.

Dias, E. 1996. Ecologia e sintaxonomia da vegetação natural dos Açores. – Ph. D. thesis, Dept de Ciências Agrárias, Universidade dos Açores.

Duelli, P. 1997. Biodiversity evaluation in agricultural landscapes: an approach at two different scales. – Agricult. Ecosyst. Environ. 62: 81–91.

Dunning, J. B., Danielson, B. J. and Pulliam, H. R. 1992. Ecological processes that affect populations in complex landscapes. – Oikos 65: 169–175.

Ebeling, F. 1978. Nordsvenska skogstyper. – Sveriges Skogsvårdsförbunds Tidskr. 4: 340–381.

Ehrlich, P. R. and Ehrlich, A. H. 1992. The value of biodiversity. – Ambio 21: 219–226.

Ellenberg, H. 1986. Vegetation ecology of central Europe. 4th ed. – Cambridge Univ. Press.

Ellenberg, H. 1996. Vegetation Mitteleuropas mit den Alpen. 5. Aufl. – Ulmer, Stuttgart.

Ellenberg, H. et al. 1991. Zeigerwerte von Pflanzen in Mitteleuropa. (Indicator values of plants in central Europe). – Scripta Geobotanica 18, Erich Goltze KG, Göttingen.

Ellenberg, Heinz and Klötzli, F. 1972. Waldgesellschaften und Waldstandorte der Schweiz. – Mitt. Schweiz. Anst. Forstl. Versuchswes. 48: 388–930.

Englemark, O. 1999. Boreal forest disturbance. – In: Walker, L. R. (ed.), Ecosystems of disturbed ground. Ecosystems of the World 16. Elsevier, pp. 161–186.

Erdelen, M. 1984. Bird communities and vegetation structure. I. Correlations and comparisons of simple and diversity indices. – Oecologia 61: 277–284.

Esseen, P.-A. et al. 1992. Boreal forest – the focal habitats of Fennoscandia. – In: Hansson, L. (ed.), Ecological principles of nature conservation. Elsevier, pp. 252–325.

Esseen, P.-A. et al. 1997. Boreal forests. – Ecol. Bull. 46: 16–47.

Esteban, J. A. et al. 1999. Research and biodiversity. A step forward. Report of an electronic conference. – Ministry of Environment, Government of Catalonia, Barcelona.

European Commission 1996. Interpretation Manual of European Union Habitats. – HAB 96/2-EN. Version EUR 15.

EU Commission 1998. Communication from the Commission to the Council and the European parliament on a forestry strategy for the European Union. – COM (1998) 649, 3/11/1998.

FAO 1999. State of the World's forests 1999. – Food and Agricultural Organisation of the United Nations, Rome, 1999.

Faraco, A. M., Fernandez, F. and Moreno, J. M. 1993. Post-fire vegetation dynamics of pine woodlands and shrublands in the Sierra de Gredos, Spain. – In: Trabaud, L. and Prodon, R. (eds), Fire in Mediterranean ecosystems, pp. 101–112.

Ferris, R. and Purdy, K. M. 1998. The application of landscape ecology theory to forest management for biodiversity in Britain: a review. – Report to Forestry Practice Division – December 1998. Forest Research, Forestry Commission, Surrey, U.K.

Ferris, R. and Humphrey, J. W. 1999. A review of potential biodiversity indicators for application in British forests. – Forestry 72: 313–328.

Fleming, R. A. 1996. A mechanistic perspective of possible influences of climate change on defoliating insects in North America's boreal forests. – In: Korpilahti, E., Kellomäki, S. and Karjalainen, T. (eds), Climate change, biodiversity and borel forest ecosystems. Silva Fenn. 30: 201–214.

Fogel, R. 1975. Insect mycophagy: a preliminary bibliography. – USDA For. Serv. Gen. Tech. Pap. Pacific Northwest Forest and Range Experiment Station.

Forman, R. T. 1995. Land mosaics. The ecology of landscapes and regions. – Cambridge Univ. Press.

Framstad, E. 1997. Vegetasjonstyper i Norge. – NINA Temahefte 12: 1–279.

Franklin, J. F. 1988. Structural and functional diversity in temperate forests. – In: Wilson, E. O. (ed.), Biodiversity. National Academy Press, Washington, D.C., pp. 166–175.

Fransson, S. 1965. The borderland. – Acta Phytogeogr. Suec. 50: 167–176.

Fries, C. et al. 1997. Silvicultural models to maintain and restore natural stand structures in Swedish boreal forests. – For. Ecol. Manage. 94: 89–103.

Fuller, R. J. and Moreton, B. D. 1989. Breeding bird populations of Kentish sweet chestnut (*Castanea sativa*) coppice in relation to the age and structure of the coppice. – J. Appl. Ecol. 24: 13–27.

Fuller, R. J. and Henderson, A. C. B. 1992. Distribution of breeding songbirds in Bradfield Woods, Suffolk, in relation to vegetation and coppice management. – Bird Study 39: 73–88.

Gaston, K. J. (ed.) 1996. Biodiversity. A biology of numbers and difference. – Blackwell.

Gaston, K. J. 1998. Biodiversity. – In: Sutherland, W. J. (ed.), Conservation science and action. Blackwell, pp. 1–19.

Godron, M. and Forman, R. T. T. 1983. Landscape modification and changing ecological characteristics. – In: Mooney, H. A. and Godron, M. (eds), Disturbance and ecosystems. Springer, pp. 12–28.

Gunnarsson, B. 1996. Bird predation and vegetation structure affecting spruce-living arthropods in a temperate forest. – J. Anim. Ecol. 65: 389–397.

Hägglund, B. and Lundmark, J.-E. 1977. Site index estimation by means of site properties of Scots pine and Norway spruce in Sweden. – Studie Forestalia Suecica 138: 1–38.

Hall, J. 1997. An analysis of National Vegetation Classification survey data. – Joint Nature Conservation Committee Report No. 272. Joint Nature Conservation Committee, Peterborough.

Halme, E. and Niemelä, J. 1993. Carabid beetles in fragments of coniferous forest. – Ann. Zool. Fenn. 30: 17–30.

Halpern, C. B. and Spies, T. A. 1995. Plant species diversity in natural and managed forests of the Pacific Northwest. – Ecol. Appl. 5: 913–934.

Hansson, L. (ed.) 1992. Ecological principles of nature conservation. – Elsevier.

Hansson, L. 1998. Indicators of biodiversity: an overview. – BEAR Tech. Rep. no. 1.

Hansson, L. 2000. Biodiversity indicators in terrestrial environments. – In: Larsson, T.-B. and Esteban, J. A. (eds), Cost effective indicators to assess biological diversity in the framework of the convention on biological diversity (CBD). Swedish Environmental Protection Agency, Stockholm, Sweden and Ministry of Environment, Government of Catalonia, Barcelona, Spain, p. 27.

Harmon, M. E. et al. 1986. Ecology of coarse woody debris in temperate ecosystems. – Adv. Ecol. Res. 15: 133–302.

Hartshorn, G. S. 1995. Ecological basis for sustainable development in tropical forests. – Annu. Rev. Ecol. Syst. 26: 155–175.

Haukioja, E. et al. 1988. The autumnal moth in Fennoscandia. – In: Berryman, A. A. (ed.), Dynamics of forest insect populations. Plenum Press, pp. 163–178.

Heikinheimo, O. 1915. The influence of shifting cultivation on forests in Finland. – Acta For. Fenn. 4.

Hobbs, R. J. 1999. Restoration of disturbed ecosystems. – In: Walker, L. R. (ed.), Ecosystems of disturbed ground. Ecosystems of the World 16. Elsevier, pp. 659–687.

Hofgaard, A. et al. (eds) 1999. Animal responses to global change in the north. – Ecol. Bull. 47.

Hofmann, G. 1994. Wälder und Forsten. Mitteleuropäische Wald- und Forst-Ökosystemtypen in Wort und Bild. (Forests. Central European forest ecosystems). – Der Wald, special issue: Waldökosystem-Katalog, Berlin.

Hörnsten, L. and Fredman, P. 2000. On the distance to recreational forests in Sweden. – Landscape and Urban Planning, in press.

Hufnagl, H. 1970. Der Waldtyp, ein Behelf für die Waldbaudiagnose. – Ried. i. I.

Huhter, V. et al. 1998. Functional implications of soil fauna diversity in boreal forests. – Appl. Soil Ecol. 10: 277–288.

Hunter, M. L. Jr 1990. Wildlife, forests and forestry. Principles of managing forests for their biological diversity. – Prentice-Hall.

Huntley, B. and Prentice, I. C. 1993. Holocene vegetation and climates of Europe. – In: Wright, Jr, H. E. et al. (eds), Global climates since the last glacial maximum. Univ. of Minnesota Press, Minneapolis, pp. 136–168.

Huse, S. 1965. Strukturformler hos urskogsbestand i øvre Pasvik. – Meld. Norges Landbrukshøgskole 44.

Huston, M. 1979. A general hypothesis of species diversity. – Am. Nat. 113: 81–101.

Huston, M. A. 1994. Biological diversity. – Cambridge Univ. Press.

James, F. C. and Wamer, N. O. 1982. Relationships between temperate forest bird communities and vegetation structure in Florida. – Ecology 63: 159–171.

Jedrzejewska, B. et al. 1994. Effects of exploitation and protection on forest structure, ungulate density and wolf predation in Bialowieza Primeval Forest, Poland. – J. Appl. Ecol. 31: 664–676.

Jennings, S. B., Brown, N. D. and Sheil, D. 1999. Assessing forest canopies and understorey illumination: canopy closure, canopy cover and other measures. – Forestry 72: 59–73.

Jensen, F. S. 1995. Forest recreation. – In: Hytönen, M. (ed.), Multiple-use forestry in the Nordic countries. METLA, The Finnish Forest Research Institute, pp. 245–278.

Johnson, C. N. 1996. Interactions between mammals and ectomycorrhizal fungi. – Trends Ecol. Evol. 11: 503–507.

Johnston, C. A., Pastor, J. and Naiman, R. J. 1993. Effects of beaver and moose on boreal forest landscapes. – In: Haines-Young, R., Green, D. R. and Cousins, S. (eds), Landscpae ecology and geographic information systems. Taylor and Francis, London, pp. 237–254.

Jonsell, M., Weslien, J. and Ehnström, B. 1998. Substrate requirements of red-listed saprozylic invertebrates in Sweden. – Biodiv. Conserv. 7: 749–764.

Järvinen, O. and Väisänen, R. 1979. Changes in bird populations as criteria of environmental change. – Holarct. Ecol. 2: 75–80.

Kaennel-Dobbertin, M. 1998a. Indicators for forest biodiversity in Europe: proposal for terms and definitions. – BEAR Tech. Rep. no. 4.

Kaennel-Dobbertin, M. 1998b. Biodiversity: a diversity in definition. – In: Bachmann, P., Kühl, M. and Päivinen, R. (eds), Assessment of biodiversity for improved forest planning. European Forest Institute Proceedings No. 18, pp. 71–81.

Kalela, A. 1961. Waldvegetatioszonen Finnlands und ihre klimatischen Paralleltypen. – Archivum Societatis Vanamo 16: 65–83.

Keller, W. et al. 1998. Waldgesellschaften der Schweiz auf floristischer Grundlage. Statistisch überarbeitete Fassung der "Waldgesellschaften und Waldstandorte der Schweiz" von Heinz Ellenberg und Frank Klötzli (1972). – Mitt. Eidgenöss. Forsch. anst. Wald Schnee Landsch. 73: 91–357.

Kimmins, J. P. 1997a. Forest ecology. A foundation for sustainable management. 2nd ed. – Prentice Hall.

Kimmins, J. P. 1997b. Certification and sustainable forestry. Will current certification initiatives deliver on their promise, or are they merely "arm-waving"? – For. Chron. 73: 166–167.

Komarkova, V. and Wielgolaski, F. E. 1999. Stress and disturbance in cold region ecosystems. – In: Walker, L. R. (ed.), Ecosystems of disturbed ground. Ecosystems of the World 16. Elsevier, pp. 39–122.

Korpel, S. 1995. Die Urwälder der Westkarpaten. – G. Fischer.

Kouki, J. and Löfman, S. 1998. Forest fragmentation: processes, concepts and implications for species. – In: Dover, J. W. and Bunce, R. G. H. (eds), Key concepts in landscape ecology. Proc. of the European Con. of The Int. Assoc. for Landscape Ecology (IALE), Myerscough College, Preston, 3–5 September 1998, pp. 187–204.

Kouki, J. and Niemelä, P. 1997. The biological heritage of Finnish forests. – In: Opus, L. L. (ed.), Finnish forests. Univ. of Joensuu, Finland, pp. 25–33.

Kuuluvainen, T. 1994. Gap disturbance, ground microtopography, and the regeneration dynamics of boreal coniferous forests in Finland: a review. – Ann. Zool. Fenn. 31: 35–51.

Lack, P. 1992. Birds on lowland farms. – HMSO, London.

Lambeck, R. J. 1997. Focal species: a multi-species umbrella for nature conservation. – Conserv. Biol. 11: 849–856.

Larsson, T.-B. and Esteban, A. (eds) 2000. Cost-effective indicators to assess biological diversity in the framework of the convention on biological diversity CBD. – Report of expert meeting in Stockholm 6–7 December 1999 and electronic conference November 1999–January 2000. Report presented at SBSTTA 5 http://www.gencat.es/mediamb/bioind/

Lawton, J. H. et al. 1998. Biodiversity inventories, indicator taxa and the effects of habitat modification in tropical forest. – Nature 391: 72–76.

Lowman, M. D. 1999. Life in the treetops. – Yale Univ. Press.

MacArthur, R. H. and MacArthur, J. W. 1961. On bird species diversity. – Ecology 42: 594–598.

Mader, H.-J. 1984. Animal habitat isolation by roads and agricultural fields. – Biol. Conserv. 29: 81–96.

Majer, A. 1968. Magyarország erdotársulásai. (Forest types of Hungary). – Akadémiai Kiadó, Budapest.

Mayer, H. 1971. Das Buchen-Naturwaldreservart Dobra/Kampleiten im niederösterreichischen Waldviertel. – Schweiz. Z. Forstwesen 122: 45–66.

Mayer, H. 1974. Wälder des Ostalpenraumes. – Gustaf Fischer, Stuttgart.

Mayer, H. 1984. Wälder Europas. – Gustav Fischer, Stuttgart.

Mayer, H. and Tichy, K. 1979. The Johannser Kogel oak *Quercus* protection area in the Zoological Garden of Lainz, Wienerwald Austria. (Das Eichen-Naturschutzgebiet Johannser Kogel im Lainzer Tiergarten, Wienerwald). – Centralbl-Gesamte-Forstwes. Wien, Osterreichischer Agrarverlag, pp. 193–226.

McGeoch, M. A. and Chown, S. L. 1998. Scaling up the value of bioindicators. – Trends Ecol. Evol. 13: 46–47.

Mikusinski, G. 1997. Woodpeckers in time and space: the role of natural and anthropogenic factors. – Acta Univ. Agricult. Suec. Silv. 23. Swedish Univ. of Agricult. Sci., Uppsala, Sweden.

Mitchell, P. L. and Kirby, K. J. 1989. Ecological effects of forestry practices in long-established woodland and their implications for nature conservation. – Oxford For. Inst. Occas. Pap. No. 39, Oxford, U.K.

Moreno, J. M., Vazquez, A. and Velez, R. 1998. Recent history of forest fires in Spain. – In: Moreno, J. M. (ed.), Large forest fires. Backhuys Publishers, Leiden, The Netherlands, pp. 159–185.

Moss, D. 1978. Song bird populations in forestry plantations. – Q. J. For. 72: 5–14.

Mucina, L., Grabherr, G. and Wallnöfer, S. 1993. Die Pflanzengesellschaften Österreichs, Bd. III, Wälder und gebüsche. – Fischer.

Murdoch, W. W., Evans, F. C. and Peterson, C. H. 1972. Diversity and pattern in plants and insects. – Ecology 53: 819–28.

Møller, P. F. 1997. Biodiversity in Danish natural forests. A comparison between unmanaged and managed forests in east Denmark. – Danmarks Geologiske Undersøgelse Rapport 1997/41.

Naiman, R. J., Melillo, J. M. and Hobbie, J. H. 1986. Ecosystem alteration of boreal forest streams by beaver (*Castor canadensis*). – Ecology 67: 1254–1269.

Naveh, Z. 1974. Effects of fire in the Mediterranean region. – In: Kozlowski, T. T. and Ahlgren, C. E. (eds), Fire and ecosystems. Academic Press, pp. 401–434.

Niemelä, J. 1999. Management in relation to disturbance in the boreal forest. – For. Ecol. Manage. 115: 127–134.

Nitare, J. and Norén, M. 1992. Nyckelbiotoper kartläggs i nytt projekt från Skogsstyrelsen. (Woodland key-habitats of rare and endangered species will be mapped in a new project of the Swedish National Board of Forestry). – Sv. Bot. Tidskr. 86: 219–226, in Swedish with English summary.

Noirfalise, A. 1984. Forêts et Stations Forestières en Belgique. – Les Presses Agronomiques de Gembloux.

Noirfalise, A. 1987. Map of the natural vegetation of the member countries of the European Community and the Council of Europe. – Office of the Official publications of the European Communities, Luxembourg.

Noss, R. F. 1990. Indicators for monitoring biodiversity: a hierarchical approach. – Conserv. Biol. 4: 355–364.

Noss, R. F. 1991. From endangered species to a biodiversity. – In: Kohm, K. (ed.), Balancing on the brink of extinction: the Endangered Species Act and lessons for the future. Island Press, pp. 227–245.

Nowicki, P., Saether, B.-E. and Solhaug, T. 1998. Threats to biodiversity. – In: Catizzone, M., Larsson, T.-B. and Svensson, L. (eds), Understanding biodiversity, a research agenda prepared by the European Working Group on Research and Biodiversity (EWGRB). Ecosystems Res. Rep. 25. European Commission, pp. 25–31.

Nygaard, B. et al. 1999. DANVEG – en database over DANske VEGetationstyper. Ver. 3. – Afdeling for Landskabsøkologi, Danmarks Miljøundersøgelser (National Environmental Research Institute).

O'Connor, R. J. and Schrubb, M. 1986. Farming and birds. – Cambridge Univ. Press.

Oldeman, L., Hakkeling, R. and Sombroeck, W. 1990. World map of the status of human-induced soil degradation: an explanatory note. – International Soil Reference and Information Center, Wageningen, the Netherlands, and United Nations Environmental Programme, Nairobi, Kenya.

Oliver, C. D. and Larson, B. C. 1990. Forest stand dynamics. – McGraw-Hill.

Östlund, L. 1993. Exploitation and structural changes in the north Swedish boreal forest 1800–1992. – Ph.D. thesis, Dept of Forest Vegetation and Ecology, Swedish Univ. of Agricult. Sci., Umeå, Sweden.

Påhlsson, L. (ed.) 1994. Vegetationstyper i Norden. – Nordiska ministerrådet, TemaNord 1994: 665.

Papanastasis, V. P. 1998. Livestock grazing in Mediterranean ecosystems: a historical and policy perspective. – In: Papanastasis, V. P. and Peter, D. (eds), Ecological basis of livestock grazing in Mediterranean ecosystems. Proceeding of the International Workshop held in Thessaloniki (Greece) 23–25 October 1997, pp. 5–9.

Parviainen, J. et al. 2000. COST Action E4. Forest reserves network in Europe. Final Report. – The Finnish Forest Research Institute, Joensuu Research Station, Finland.

Pearson, D. L. and Cassola, F. 1992. World-wide species richness of tiger beetles (Coleoptera: Cicindelidae): indicator taxon for biodiversity and conservation studies. – Conserv. Biol. 6: 376–391.

PEFCC 1999. Pan European Forest Certification Framework. Common elements and requirements. Technical document. – As adopted by the PEFCC General Assembly on the 30th June 1999, Paris, France.

Pérez, B. and Moreno, J. M. 1998. Fire-type and forestry management effects on the early postfire vegetation dynamics of a *Pinus pinaster* woodland. – Plant Ecol. 134: 27–41.

Perry, D. A. 1994. Forest ecosystems. – The John Hopkins Univ. Press.

Peterken, G. 1993. Woodland conservation and management. 2nd ed. – Chapman and Hall.

Peterken, G. F. 1996. Natural woodland: ecology and conservation of northern temperate regions. – Cambridge Univ. Press.

Peterken, G. P. 1981. Woodland conservation and management. – Chapman and Hall.

Petraitis, P. S., Latham, R. E. and Niesenbaum, R. A. 1989. The maintenance of species diversity by disturbance. – Q. Rev. Biol. 64: 393–418.

Pickett, S. T. A. and White, P. S. 1985. The ecology of natural disturbance and patch dynamics. – Academic Press.

Pignatti, S. 1998. I boschi d'Italia. – Sinecologia e biodiversità. UTET.

Pimentel, D. (ed.) 1993. World soil erosion and concervation. – Cambridge Univ. Press.

Pimentel, D. and Harvey, C. 1999. Ecological effects of erosion. – In: Walker, L. R. (ed.), Ecosystems of disturbed ground. Ecosystems of the World 16. Elsevier, pp. 123–135.

Platt, W. J. and Strong, D. M. (eds) 1989. Special feature: treefall gaps and forest dynamics. Ecology 70: 536–576.

Pott, R. 1995. Die Pflanzengesellschaften Deutschlands. (The plant associations of Germany). 2nd ed. – Ulmer, Stuttgart.

Prendergast, J. R. et al. 1993. Rare species, the coincidence of diversity hotspots and conservation strategies. – Nature 365: 335–337.

Quine, C. P., Humphrey, J. W. and Ferris, R. 1999. Should wind disturbance patterns observed in natural forests be mimicked in planted forests in the British uplands? – Forestry 72: 337–358.

Rackham, O. and Moody, J. 1996. The making of the Cretan landscape. – Manchester Univ. Press.

Raivio, S. 1988. The peninsula effect and habitat structure: bird communities in coniferous forests of the Hanko Peninsula, southern Finland. – Ornis Fenn. 65: 129–149.

Rameau, J. C. 1997. Nomenclature, CORINE Biotopes, Types d'habitats français, Forêts. – Ecole nationale du genie rural des eaux et forets, Nancy, pp. 97–144.

Ratcliffe, P. R. 1993. Biodiversity in Britain's forests. – The Forestry Authority, Forestry Commission.

Rego, F. and Dias, S. 1999. Indicators for monitoring and evaluation of forest biodiversity in Europe – aims, tasks and the significance of the BEAR project for pan-european nature conservation. – In: Bischoff, C. and Dröschmeister, R. (eds), European monitoring for nature conservation. Bonn-Bad Godesberg (Bundesamt für Naturschutz), Schriftenr. Landschaftspflege Naturschutz 62.

Reid, M. V. et al. 1993. Biodiversity indicators for policy makers. – World Resources Institute, Washington D.C., USA.

Rivas-Martínez, S. 1987. Mapa de series de vegetación de España 1:400000. – Instituto Nacional para la Conservación de la Naturaleza (ICONA), Madrid.

Rivas-Martínez, S. et al. 1994. El proyecto de cartografía e inventariación de los tipos de hábitats de la Directiva 92/43/CEE en España. – Colloq. Phytosociol. 22: 611–661.

Rivas-Martínez, S., Fernández-González, F. and Loidi, J. 1999. Checklist of plant communities of Iberian Peninsula, Balearic and Canary Islands to suballiance level. – Itinera Geobot. 12, in press.

Rodwell, J. S. (ed.) 1991. British plant communities. Vol. 1. Woodlands and scrub. – Cambridge Univ. Press.

Rose, F. 1992. Temperate forest management: its effects on bryophyte and lichen floras and habitats. – In: Bates, J. W. and Farmer, A. M. (eds), Bryophytes and lichens in a changing environment. Clarendon Press, pp. 211–233.

Rosenzweig, M. L. 1995. Species diversity in space and time. – Cambridge Univ. Press.

Ruiz de la Torre, J. 1990. Mapa Forestal de España Escala 1:200.000. – Memoria General. Instituto Nacional para la Conservación de la Naturaleza (ICONA). Ministerio de Agricultura, Pesca y Alimentación, Madrid.

Sage, R. B. and Robertson, P. A. 1996. Factors affecting songbird communities using new short rotation coppice habitats in spring. – Bird Study 43: 201–213.

Samuelsson, J., Gustafsson, L. and Ingelög, T. 1994. Dying and dead trees. A review of their importance for biodiversity. – Swedish Threatened Species Unit, Uppsala, Sweden.

Santos, A. 1990. Bosques de Laurisilva en la region Macaronésica. – Council of Europe, Strasbourg.

Schmidt, P. A. 1995. Übersicht der natürlichen Waldgesellschaften Deutschlands. (Overview of the natural forest associations in Germany). – Schriftenr. Sächs. LA Forsten 4, Graupa.

Schowalter, T. D. and Lowman, M. D. 1999. Forest herbivory: insects. – In: Walker, L. R. (ed.), Ecosystems of disturbed ground. Ecosystems of the World 16. Elsevier, pp. 253–269.

Schulze, E.-D. and Mooney, H. A. (eds) 1994. Biodiversity and ecosystem function. – Springer.

Sjöberg, K. and Lennartsson, T. 1995. Fauna and flora management in forestry. – In: Hytönen, M. (ed.), Multiple-use forestry in the Nordic countries. METLA, The Finnish Forest Resarch Institute. Gummerus Printing, Jyväskylä, pp. 191–243.

Sjögren, E. 1972. Vascular plant communities of Madeira. – Bol. Mus. Munic. Do Funchal XXVI: 45–125.

Soó, R. 1960. Magyarország erdőtársulásainak és erdőtípusainak áttekintése. (Forest associations and forest types of Hungary). – Az Erdo 9: 321–340.

Soó, R. 1964, 1966, 1968, 1970, 1973, 1980. A magyar flóra és vegetáció rendszertani-növényföldrajzi kézikönyve I–VI. (Taxonomic and phytogeographic handbook of the Hungarian vegetation I–VI). – Akadémiai Kiadó, Budapest.

Spanos, K. A. et al. 1998. Classification of forest vegetation in Greece. – BEAR Tech. Rep. no. 3.

Speight, M. C. D. 1989. Saproxylic invertebrates and their conservation. – Council of Europe, Nature and Environment Ser. 42.

Speth, J. G. 1994. Towards an effective and operational international convention on desertification. United Nations International Convention of Desertification. – International Negotiating Committee, New York City.

Spies, T. 1997. Forest stand structure, composition, and function. – In: Kohm, A. and Franklin, J. F. (eds), Creating a forestry for the 21st Century. The science of ecosystem management, pp. 11–30.

Staines, B. et al. 1998. Desk and limited field studies to analyse the major factor influencing regional deer populations and ranging behaviour. – Final project report to the Ministry of Agriculture, Fisheries and Food, U.K. MAFF Project VC 0314.

Standovár, T. 1996. Aspects of diversity in forest vegetation. – In: Bachmann, P., Kuusela, K. and Uutera, J. (eds), Assessment of biodiversity for improved forest management. EFI Proc., 6. European Forestry Inst., Joensuu.

Starlinger, F. 2000. Vegetationskundliche Charakterisierung von sekundären Nadelwäldern und Nadelholz-Forsten. – FBVA-Berichte, Wien 111: 2–29.

Terradas, J. 1999. Holm oak and holm oak forests: an introduction. – Ecol. Stud. 137: 1–13.

Thirwood, J. V. 1981. Man and the Mediterranean forest. – Academic Press.

Tomter, S. M. 1994. Skog 1994. Statistics of forest conditions and resources in Norway. – NIJOS report.

Tsoumis. G. 1985. The depletion of forests in the Mediterranean region – a historical review from the ancient times to the present. – Scientific annals of the Dept of Forestry and Natural Environment, Vol. KH 11: 281–300.

Tüxen, R. 1956. Die heutige potentielle Vegetation als Gegenstand der Vegetationskartierung. – Angewandte Pflanzensoz. 13: 5–32.

Van Bueren, E. and Blom, E. M. 1997. Hierarchical framework for the formation of sustainable forest management standards. – The Tropenbos Foundation, Leiden.

van der Werf, S. 1991. Natuurbeheer in Nederland. Deel 5: Bosgemeenschappen. – Pudoc, Wageningen.

Vasander, H. (ed.) 1996. Peatlands in Finland. – Finnish Peatland Society.

Walker, B. H. 1992. Biodiversity and ecological redundancy. – Conserv. Biol. 6: 18–23.

Whitbread, A. M. and Kirby, K. J. 1992. Summary of national vegetation classification woodland descriptions. – U.K. Nature Conservation No. 2. Joint Nature Conservation Committee, Peterborough.

White, K. A. and Crist, T. O. 1995. Critical thresholds in species' responses to landscape structure. – Ecology 76: 2446–2459.

Wiens, J. A. 1990. Habitat fragmentation and wildlife populations: the importance of autecology, time, and landscape structure. – Trans. 19th IUGB Congress, Trondheim, pp. 381–391.

Wiens, J. A. 1995. Habitat fragmentation: island versus landscape perspectives on bird conservation. – Ibis 137: 97–104.

Wiens, J. A. et al. 1993. Ecological mechanisms and landscape ecology. – Oikos 66: 369–380.

Wilcove, D. S., McLellan, C. H. and Dobson, A. P. 1986. Habitat fragmentation in the temperate zone. – In: Soulé, M. E. (ed.), Conservation biology. The science of scarcity and diversity. Sinauer, pp. 237–256.

Wilmans, O. 1998. Ökologische Pflanzensoziologie. (Ecological plant sociology). 6th ed. – Verlag Quelle & Meyer, Wiesbaden.

Zackrisson, O. 1977. Influence of forest fires on the north Swedish boreal forest. – Oikos 29: 22–32.

Zipper, W. C. 1993. Deforestation patterns and their effects on forest patches. – Landscape Ecol. 8: 177–184.

Appendix 1.

BEAR partners and user-panel

The 27 partners of the EU project "Indicators for monitoring and evaluation of forest biodiversity in Europe" (Contract FAIR-CT97-3575)

Swedish Environmental Protection Agency

Research Department
SE-106 48 Stockholm
Sweden
Tor-Björn Larsson
Tel +46 8 6981447
Fax +46 8 6981663
e-mail: tor-bjorn.larsson@environ.se

Swedish University of Agricultural Sciences

Grimsö Wildlife Research Station
Department of Conservation Biology
SE-730 91 Riddarhyttan
Sweden
Per Angelstam
Tel +46 581 69 73 02
Fax +46 581 69 73 10
e-mail: per.angelstam@nvb.slu.se

The Forestry Research Institute of Sweden

SkogForsk
Glunten
SE-751 83 Uppsala
Sweden
Lena Gustafsson
Tel +46 18 188500
Fax +46 18 188600
e-mail: lena.gustafsson@skogforsk.se

Geological Survey of Denmark and Greenland

Thoravej 8
DK-2400 Copenhagen NV
Denmark
Richard Bradshaw, Peter Friis Møller
Tel +45 38142350
Fax +45 38142050
e-mail: rhwb@geus.dk

Norwegian Institute for Nature Research

Tungasletta 2
N-7005 Trondheim
Norway
Bjørn Åge Tømmerås
Tel +47 73 80 15 52
Fax +47 73 80 14 01
e-mail: bjorn.a.tommeras@ninatrd.ninaniku.no

University College Cork

Department of Zoology and Animal Ecology
Lee Maltings
Prospect Row
Cork
Ireland
John O'Halloran
Tel +353 21 904193
Fax +353 21 270562
e-mail: johalloran@ucc.ie

Centre for Ecology and Hydrology

Hill of Brathens, Glassel
Banchory
Scotland
U.K. AB31 4BY
Allan D. Watt
Tel +44 1330 826300
Fax +44 1330 823303
e-mail: adw@ceh.ac.uk

Forestry Commission Research Agency

Alice Holt Lodge
Wrecclesham
Farnham, Surrey
U.K. GU10 4LH
Richard Ferris
Tel +44 1420 22255
Fax +44 1420 23653
e-mail: r.ferris@forestry.gov.uk

Wageningen Univ. and Research Center

Dept of Forest Environmental Sciences
Silviculture and Forest Ecology Group
P.O. Box 342
NL-6700 AH Wageningen
The Netherlands
G. M. J. (Frits) Mohren
Tel +31 317 482926
Fax +31 317 483542
e-mail: frits.mohren@btbo.bosb.wau.nl

In cooperation with:

Alterra, Green World Research

P.O. Box 47
NL-6700 AA Wageningen
The Netherlands
Rienk-Jan Bijlsma
Tel +31 317 477906
Fax +31 317 419000
e-mail: r.j.bijlsma@alterra.wag-ur.nl

Instituut Voor Bosbouw En Wildbeheer

Gavestraat 4
B-9500 Geraardsbergen
Belgium
Kris Vandekerkhove, Diego VanDenMeersschaut
Tel +32 54 437111
Fax +32 54 4110896
e-mail: kkandekerkhove@ibw.be

Laboratoire d'Entomologie Forestière, INRA

Station de Recherches Forestières
Pierroton, BP 45
F-33611 Gazinet Cedex
France
Hervé Jactel
Tel +33 5 57979027
Fax +33 5 57979088
e-mail: jactel@pierroton.inra.fr

Universität Göttingen

Fakultät für Forstwissenschaften und Waldökologie
Institut für Forstpolitik, Forstgeschichte und Naturschutz
Arbeitsgebiet Naturschutz
Büsgenweg 5
D-37077 Göttingen
Germany
Klaus Halbritter
Tel +49 551 39 3412 (secretary)
Fax +49 551 39 3415
e-mail: khalbri@gwdg.de

Forstliche Versuchs- und Forschungsanstalt Baden-Württemberg

Abteilung Botanik und Standortskunde
P.O. Box 708
D-79007 Freiburg
Germany
Winfried Bücking
Fax +49 761 4018 333
e-mail: buecking@ruf.uni-freiburg.de

Federal Research Centre for forestry and Forest Products

Leuschnerstrasse 91
D-21031 Hamburg
Germany
Hermann Ellenberg
Tel +49 40 73962 113
Fax +49 40 73962 480
e-mail: ellenberg@aixh0001.holz.uni-hamburg.de

Forstliche Bundesversuchsanstalt

Institut für Waldbau
Hauptstrasse 7
A-1140 Wien
Austria
Georg Frank
Tel +43 1 87838 2208
Fax +43 1 87838 2250
e-mail: fbva@forvie.ac.at

Estação Florestal Nacional and Centro de Ecologia Aplicada "Prof. Baeta Neves" Instituto Superior de Agronomia

Tapada da Ajuda
PT-1349-017 Lisboa
Portugal
Francisco Rego, Susana Dias
Tel +351 21 3616080
Fax +351 21 3623493
e-mail: ceabn@ip.pt

Jardin Botanico da Madeira

Quinta do bom-sucesso
PT-9050 Funchal
Madeira
Francisco Manuel Fernandes
Tel +351 91 2002027
Fax +351 91 2002006
e-mail: goodyera@dragoeiro.uma.pt

Facultad de Ciencias del Medio Ambiente, Universidad de Castilla-La Mancha

E-45071 Toledo
Spain
Frederico Fernández-Gonzalez
Tel +34 925 268800 ext. 5417
Fax +34 925 268840
e-mail: ffernan@amb-to.uclm.es

Academia Italiana di Science Forestali

Piazza Edison 11
I-50133 Firenze
Italy
Orazio Ciancio, Marco Marchetti
Tel +39 055 570348
Fax +39 055 575724
e-mail: ciancio@unifiit.it

National Agricultural Research Foundation NARF

Forest Research Institute
GR-57006 Vassilika, Thessaloniki
Greece
Konstantinos Spanos
Tel +30 31 461 171
Fax +30 31 461 341
e-mail: kspanos@fri.gr

European Forest Institute

Torikatu 34
FIN-80100 Joensuu
Finland
Janne Uuttera
Tel +358 13 252020
Fax +358 13 124393
e-mail: efisec@efi.fi

University of Joensuu

Faculty of Forestry
P.O. Box 111
FIN-80101 Joensuu
Finland
Pekka Niemelä
Tel +358 13 2514407
Fax +358 13 2514444
e-mail: pekka.niemela@forest.joensuu.fi

Universität für Bodenkultur

Institute of Forest Economics
Gregor Mendel Str. 33
A-1180 Vienna
Austria
Ewald Rametsteiner
Tel +43 1 47 654 4403
Fax +43 1 47 654 4407
e-mail: ramet@edv1.boku.ac.at

Institut National de la Recherche Agronomique (INRA), SAD Toulose, UMR 5552 Laboratoire d'Ecologie Terrestre

Chemin de Borde Rouge
BP 27
F-31326 Castanet-Tolosan Cedex,
France
Gérard Balent
Tel +33 5 61 28 52 58
Fax +33 5 61 73 20 77
e-mail: balent@toulouse.inra.fr

Swiss Federal Institute for Forest, Snow and Landscape Research

Zuercherstrasse 111
CH-8903 Birmensdorf
Switzerland
Christoph Scheidegger, John Innes
Tel +41 1 739 2439
Fax +41 1 739 2215
e-mail: christoph.scheidegger@wsl.ch

Eötvös University

Department of Plant Taxonomy and Ecology
Ludovika tér 2
H-1083 Budapest
Hungary
Tibor Standovár
Tel +36 1 210 1084
Fax +36 1 333 8764
e-mail: h7184sta@ella.hu, standy@ludens.elte.hu

University of Ljubljana

Biotechnical Faculty, Department of Forestry
Vecna pot 83
1000 Ljubljana
Slovenia
Jurij Diaci, Andrej Boncina
Tel +386 61 123 11 61
Fax +386 61 27 11 69
e-mail: diaci.jurij@uni-lj.si

BEAR user-panel of organisations that have agreed to comment upon preliminary results

1. Confederation of European Forest Owners

Rue des Fripiers 17
Galerie du Centre, Bloc 2, 5e
B-1000 Brussels
Belgium
Contact person: Natalie Hufnagl, Secretary General
Tel +32 22 19 02 31
Fax +32 22 19 21 91
e-mail: cepf@planetinternet.be

2. WWF International

Avenue du Mont-Blanc
CH-1196 Gland
Switzerland
Contact person: Per Rosenberg, Forest Officer
Tel +41 22 3649229
Fax +41 22 3643239
e-mail: prosenberg@wwfnet.org

3. National Board of Forestry

Environmental Department
SE-551 83 Jönköping
Sweden
Contact person: Erik Sandström, Senior Advisor
Tel +46 36155714
Fax +46 36166170
e-mail: erik.sandstrom@svo.se

4. The Swedish Federation of Forest Owners

SE-105 33 Stockholm
Sweden
Contact person: Sven Sjunnesson, Forest Director
Tel +46 8 787 5400
Fax +46 8 787 5908
e-mail: riks@skogsagarna.se

5. Timber Growers Association TGA

5 Dublin Street Lane South
Edinburgh
U.K. EH1 3PX
Contact person: Ben Gunneberg, Technical Director
Tel +44 131 538 7111
Fax +44 131 538 7222
e-mail: ben@timber-growers.co.uk

6. Office National des Forêts

Departement des Recherches Techniques
Boulevard de Constance
F-77300 Fontainebleau
France
Contact person: M. Vallance, Directeur de Recherches
Tel +33 1 60749228
Fax +33 1 64224973
e-mail: dtc5onf@calvanet.calvacom.fr

7. Union des Sylviculteurs du Sud de l'Europe USSE

Av. Madariaga, 1-1-DTO 6
E-48014 Bilbao-Bizkaya
Spain
Contact person: J. L. Martres, Le Président du Directoire
Tel +34 4420 6990
Fax +34 4476 3715
e-mail: usse@jet.es

information also to the French correspondent

Maison de la Foret
6, parvis des Chartrons
F-33075 Bordeaux Cedex
France
Contact person: Dominique Merzeau

8. Niedersächsiches Forstplanungsamt

Forstweg 1 A
D-38302 Wolfenbüttel
Germany
Contact person: Christian Weigel
Tel +49 5331 30030
Fax +49 5331 300379

Alpine and subalpine coniferous forests

42.21 – 42.23	Acidophilous forests (Vaccinio-Piceetea)
42.31 and 42.32	Alpine forests with larch and *Pinus cembra*
42.4	*Pinus uncinata* forests (* on gypsum or limestone)

Mediterranean mountainous coniferous forests

42.14	*Appenine *Abies alba* and *Picea excelsa* forests
42.19	*Abies pinsapo* forests
42.61 – 42.66	*Mediterranean pine forests with endemic black pines
42.8	Mediterranean pine forests with endemic Mesogean pines, including *Pinus mugo* and *Pinus leucodermis*
42.9	Macaronesian pine forests (endemic)
42.A2 – 42.A5	*Endemic Mediterranean forests with *Juniperus* spp. and 42.A8
42.A6	*Tetraclinis articulata* forests (Andalusia)
42.A71 – 42.A73	*Taxus baccata* woods

Other relevant habitat types

Coastal sand dunes and continental dunes

16.29	Wooded dunes of the Atlantic coast
16.27	*Dune juniper thickets (*Juniperus* spp.)
16.29 × 42.8	*Wooded dunes with *Pinus pinea* and/or *Pinus pinaster*

Temperate heath and scrub

5.5	*Scrub with *Pinus mugo* and *Rhododendron hirsutum* (Mugo-Rbododenretum hirsuti)
31.622	Sub-Arctic willow scrub

Sclerophyllous scrub (matorral)

31.88	*Juniperus communis* formations on calcareous heaths or grasslands
32.131 – 32.13	Juniper formations
32.17	*Matorral with *Zyziphus*
32.18	*Matorral with *Laurus nobilis*
32.216	Laurel thickets

Natural and semi-natural grassland formations

32.11	Sclerophyllous grazed forests (dehesas) with *Quercus suber* and/or *Quercus ilex*

*= The sign indicates priority habitat types

Appendix 3.

CORINE land cover types of relevance to forest biodiversity

Level 1 distinguishes the following five categories:
1. Artificial surfaces;
2. Agricultural areas;
3. Forest and semi-natural areas;
4. Wetlands;
5. Waterbodies.

For the purposes of developing Biodiversity Evaluation Tools for European forests we are interested in further sub-dividing Forest and semi-natural areas according to level 2 in:

3.1 Forest
3.2 Scrub and/or herbaceous vegetation associations
3.3 Open areas with little or no vegetation

Finally, these were divided into level 3 categories that are of greater interest for forest biodiversity and that have correspondence to the broad groups of Forest Types for Biodiversity Assessment defined for BEAR (cf. Chapter 3.2).

3.1 Forest

3.1.1 Broad-leaved forest

This category typically includes forest where native deciduous broad-leaved species dominate (*Fagus sylvatica, Quercus robur, Q. petraea, Q. pyrenaica, Q. pubescens, Q. cerris, Carpinus betulus, Fraxinus excelsior, F. angustifolia, Acer pseudoplatanus, A. platanoides, Tilia cordata, T. platyphyllos, Ostrya carpinifolia, Castanea sativa, Betula pendula, B. pubescens, B. celtiberica, Alnus glutinosa, Populus tremula, Sorbus aucuparia, Ulmus minor* and *U. glabra*). This corresponds to part of Bohn's formations F2 (mesophytic deciduous broad-leaved forests) and G (thermophilous deciduous broad-leaved forests).

3.1.2 Coniferous forests

Formations dominated by native conifers (*Abies alba, A. cephalonica, Picea abies, Larix decidua, Pinus cembra, Pinus uncinata, P. mugo, P. nigra, P. sylvestris, P. pinaster, P. pinea, P. halepensis, Cupressus sempervirens, Taxus baccata,*

Juniperus thurifera, J. oxycedrus and *J. phoenicea*). This corresponds partly to Bohn's groups D (mesophytic and hygronesophytic coniferous forests) and K (Xerophytic coniferous forests, woodlands and scrub).

3.1.3 Mixed forest

Forest where broad-leaved and coniferous species co-dominate (a mixture of the two groups above). This mixed forest category can be present in Bohn's groups D, F, G, and H.

3.2 Scrub and/or herbaceous vegetation associations

3.2.1 Natural grasslands

Areas often situated on rough uneven ground, frequently including rocky areas. This can correspond to Bohn's formations M (steppes) and N (oroxerophytic vegetation) but are of limited interest for discussion under BEAR.

3.2.2 Moors and heathland

Areas of low cover, dominated by bushes, shrubs *Ulex* spp., *Erica* spp., *Calluna vulgaris* and herbaceous plants, often corresponding to tundra or alpine vegetation (b), subarctic, boreal and nemoral-montane (woodlands and subalpine vegetation (c) or Atlantic dwarf shrub-heaths (E).

3.2.3 Sclerophyllous vegetation

Includes the discontinuous bushy associations of Mediterranean garrigues and the dense shrub formations of the maquis. It includes species as *Quercus suber, Q. ilex, Q. coccifera, Olea europea* ssp. *sylvestris, Ceratonia siliqua, Pistacia lentiscus,* and *Myrtus communis*. This category is equivalent to Bohn's Type J (Mediterranean broad-leaved sclerophyllous forest and shrub).

3.2.4 Transitional woodland/shrub

Bushy or herbaceous vegetation with scattered trees. This is defined as a transitional stage between categories, with no equivalent in Bohn's scheme. However, it can be locally very important as in the Portuguese "montados" or the Spanish "dehesas".

Potential natural vegetation of Europe

Types identified in an international project for the construction of a map of the potential natural vegetation of Europe at a scale of 1: 2.5 million (Bohn 1994, 1995). On a European scale the suggested BEAR Forest Types for Biodiversity Assessment FTBAs correspond to one or several of the types below of potential natural vegetation, cf. Chapter 3.2.

Zonal and extrazonal vegetation (depending primarily on climate):

Main formations	First division of the main formation unit according to the species composition of the tree layer depending on climate and soil conditions
A Polar deserts and subnival vegetation of high mountains	A1 Polar deserts A2 Subnival vegetation of high mountains in the boreal and nemoral zone
B Tundras and alpine vegetation	B1 Artic tundras B2 Northern tundras B3 Southern tundras B4 Mountain tundras and sparse orarctic vegetation B5 Alpine vegetation in the boreal, nemoral and Mediterranean zone
C Subarctic, boreal and nemoral-montane woodlands and subalpine vegetation	C1 Subarctic open woodlands (*Betula czerepanovii, Picea obovata, Pinus sylvestris*) C2 West-boreal and nemoral-montane birchforests, partly with *Pinus* (*Betula czerepanovii, B. pubescens, Pinus sylvestris*) C3 Subalpine vegetation (forests, shrub and dwarf shrub communities in combination with grasslands and tall-herb communities) in the nemoral and Mediterranean zone
D Mesophytic and hygromesophytic coniferous and broad-leaved coniferous forests	West boreal spruce forests (*Picea abies, P. obovata*), partly with pine, birch (*Pinus sylvestris, Betula czerepanovii*): D1 North boreal types D2 Middle boreal types D3 South boreal types East boreal pine-spruce- and fir-spruce forests (*Picea obovata, Pinus sibirica, Abies sibirica*), partly with birch (*Betula czerepanovii*), larch (*Larix sibirica*): D4 North boreal type D5 Middle boreal types D6 South boreal types D7 Montane (Ural) types

D8 Hemiboreal spruce and fir-spruce-forests (*Picea abies, P. obovata, Abies sibirica*) with broad-leaved trees (e.g. *Quercus robur, Tilia cordata, Ulmus glabra, Acer platanoides*)

D9 Montane-altimontane, partly submontane fir- and spruce forests (*Abies alba, A. nordmanniana, A. borisii-regis, Picea abies, P. omorica, P. orientalis*) in the nemoral zone
Boreal and hemiboreal pine forests (*Pinus sylvestris*), partly with birch, spruce (*Betula czerepanovii, Picea obovata, P. abies*)

D10 North boreal pine forests

D11 Middle and south to hemiboreal pine forests

D12 Hemiboreal and nemoral pine forests (*Pinus sylvestris, P. peuce, P. heldreichii, P. kochiana*) partly with broad-leaved trees

E Atlantic dwarf shrub heaths

E1 Boreo-atlantic heaths

E2 North-atlantic aerohaline heaths

E3 Middle-atlantic aerohaline heaths

E4 South-atlantic aerohaline heaths

F Mesophytic deciduous broad-leaved and coniferous broad-leaved forests

F1 Oak- and mixed oak forests, poor in species (*Quercus robur, Q. petraea, Q. pyrenaica, Pinus sylvestris, Betula pendula, B. pubescens, B. celtiberica*)

F2 Mixed oak-ash forests (*Fraxinus excelsior, F. angustifolia, Quercus robur, Ulmus glabra, Quercus petraea*)

F3 Mixed oak-hornbeam forests (*Carpinus betulus, Quercus robur, Q. petraea, Tilia cordata*)

F4 Mixed lime-oak forests

F5 Beech and mixed beech forests

F6 Oriental beech and hornbeam-oriental beech forests (*Fagus orientalis, Carpinus betulus, C. caucasica*)

F7 Mixed Caucasian hornbeam-oak forests (*Quercus robur, Q. petraea, Carpinus caucasica*)

G Thermophilous deciduous broad-leaved forests and mixed coniferous broad-leaved forests

G1 Subcontinental, partly with seasonal changes in humidity, partly acidophilous oak- and pine-oak forests (*Quercus robur, Q. petraea, Q. dalechampii, Q. pubescens, Pinus sylvestris*)

G2 Pannonic-subcontinental mixed maple-oak-steppe woodlands (*Quercus robur, Q. petraea, Q. pendunculiflora, Q. pubescens, Q. virginiana, Q. cerris, Acer tataricum, A. campestre*)

G3 Subcontinental-Submediterranean and Supramediterranean oak-forests (*Quercus petrea, Q. dalechampii, Q. polycarpa, Q. cerris, Q. frainetto*)

G4 Submediterranean and Supramediterranean mixed oak-forests and oak dominated scrub-forests (*Quercus pubescens, Q. faginea, Q. broteroi, Q. canariensis, Q. pyrenaica, Q. brachyphylla, Fraxinus ornus, Ostrya carpinifolia, Carpinus orientalis, Castanea sativa*)

H Humid thermophytic mixed broad-leaved forests

J Mediterranean broad-leaved sclerophyllous forests and shrub	Meso- and Supramediterranean and relictic sclerophyllous forests:
	J1 *Quercus rotundifolia* forests
	J2 Holm oak forests (*Quercus ilex*)
	J3 Cork oak forests (*Quercus suber*)
	J4 Kermes oak forests and shrubs (*Quercus coccifera*)
	Thermomediterranean sclerophyllic forests and xerophytic scrubs:
	J5 Thermomediterranean cork oak forests (*Quercus suber*)
	J6 Thermomediterranean *Quercus rotundifolia* forests
	J7 Wild olive-locust tree forests (*Ceratonia siliqua, Olea europaea*)
	J8 Thermomediterranean xerophytic scrub
K Xerophytic coniferous forests, woodlands and shrub	K1 Nemoral, sub- and Oromediterranean pine-forests (*Pinus sylvestris, P. nigra, P. heldreichii*)
	K2 Meso- to Thermomediterranean pine forests (*Pinus pinea, P. halepensis, P. brutes, P. pityusa*)
	K3 Meso- and Supramediterranean fir forests (*Abies pinsapo, A. cephalonica*)
	K4 Juniper and cypress forest and scrub (*Juniperus thurifera, J. excelsa, J. foetidissima, J. polycarpos, Cupressus sempervirens*)
L Forest steppes (meadow steppes, alternating with deciduous broad-leaved forests)	L1 Northern (subcontinental) meadow steppes alterning with oak-forests (*Quercus robur*)
	L2 Southern (Submediterranean) lowland-colline herb-grass steppes alternating with oak forests (*Quercus pubescens, Q. robur, Q. pendunculiflora*) with *Acer tataricum*
M Steppes	M1 – M4
N Oroxerophytic vegetation (thorn-cushion communities, mountain steppes)	
O Deserts	

Azonal vegetation (depending on soil and hydrological conditions)

Main formations	**First division of the main formation unit according to the species composition of the tree layer depending on climate and soil conditions**
P Coastal and inland halophytic vegetation	P1 – P2
R Reed and sedge swamps	
S Mires	S1 Ombrotrophic mires
	S2 Arctic-subarctic ombro-minerotrophic mires
	S3 Minerotrophic mires (incl. swamp forests)
T Swamp and fen forests	T1 Carrs with *Alnus glutinosa*
	T2 Carrs with *Betula pubescens*

U Vegetation of flood plains, estuaries and fresh water polders

U1 Subarctic flood plain shrubs
U2 Boreal flood plain forests
U3 Flood plain forests and fresh water polders of the nemoral zone
U4 Mediterranean flood plain forests (*Platanus orientalis, Fraxinus angustifolia, Phoenix theophrasti, Nerium oleander, Tamarix* spp.)
U5 Continental tamarisk-flood plain shrubs
U6 Flood-plain meadows in complex with flood plain forests and shrubs

Appendix 5.

European forest types schemes

There exist several schemes of forest types. Such schemes were developed for practical use (in forestry) and/or scientific purposes (forestry, vegetation mapping, landscape analysis etc.) in most European countries. Each system has been designed to serve a certain purpose and/or is based upon the tradition of a certain branch of science.

Below is an overview of most currently used and/or important forest type schemes in Europe.

Finland

Cajander's forest types: used in practical forest planning for estimating forest productivity. Based on ground layer vegetation communities and moisture. Six major forest types and their parallel types are used for four vegetation zones (Cajander 1949, Kalela 1961). A separate classification system is used for peatlands. Classification is based on water regime, nutrient level and the amount and composition of the tree stock. Four major types, ca 30 types separated based on water regime and nutrient level (e.g. Vasander 1996).

Sweden

The North Swedish Forest Type Scheme (Arnborg 1964) should first be mentioned for the boreal forest. This scheme is a 4 × 4 site classification table based on soil moisture (4 groups: wet-dry) and field vegetation types reflecting nutrient conditions (4 types: poor-rich). It was originally established to identify suitable regeneration measures. However, it got a very widespread use in practical forestry and was modified and improved by e.g. Arnborg (1964) and Ebeling (1978). One version developed by Tamm and Holmen (1961) was employed by the Swedish Forestry Survey. It was replaced from ca 1975 and onwards by a system that more directly predicted the potential productivity of wood (Hägglund and Lundmark 1977). In the latter, site classification was based on field layer vegetation and moisture. Ecologists have developed a quite different classification system based to a large extent on the communities of field layer plants (Anon. 1984) but it has been of limited use in both applied and basic forest science.

Norway

The National Forest Inventory (NFI) is a sample plot inventory with the aim of providing data on natural resources and environment for forest land in Norway. The following classification of the productive forest areas is used: lichen forest, cowberry forest, bog whortleberry forest, bilberry forest, small fern forest, small herb forest, tall fern forest, tall forb forest, rich deciduous forest, swamp forest, ombrotophic bog forest and *Calluna* heath (Tomter 1994). The system is comparable to the criteria established for all types of forest land, including unproductive forests, developed by Fremstad (1997).

Ireland

No single habitat classification scheme is widely used in Ireland at present, but a "Standard Approach to Habitat Classification in Ireland" has recently been drafted (Anon. 1999). This first-step approach is aimed at widespread, non-specialist use, and includes 17 categories of woodland and scrub (5 categories of semi-natural woodland, 4 of other woodland, 3 of low forest (scrub), 2 of linear woodland features, and 3 of miscellaneous). Specialist studies of Irish woodlands have also attempted to classify Irish woodlands in greater detail by reference to other European schemes (e.g. Braun-Blanquet and Tüxen 1952, Rodwell 1991).

U.K.

There are two "forest type" classification schemes that are most commonly used in the U.K. The first is the National Vegetation Classification (NVC) which aims to describe the whole range of British vegetation as a series of plant communities. The woodland section (Vol. 1, Rodwell 1991) comprises 25 communities; seven types of wet woodland, eleven dry-land high forest communities and seven scrubs and underscrubs (Rodwell 1991). Summary descriptions of the communities (excluding shrubs), and an assessment of their relationship to each other are provided by Whitbread and Kirby (1992), and maps of the distribution of these forest types published by Hall (1997).

The second classification scheme is the Stand Type System (Peterken 1981), which divides woodland into 12 stand groups and 39 stand types largely on the basis of the trees and shrubs, but also using some soil characters. This scheme was has largely been replaced by the NVC.

Belgium

Classification based on "Noirfalise" forest types. This system classifies 32 forest types with respect to vegetation and soil characteristics, but especially concentrates on the Walloon part (southern part) of Belgium (Noirfalise 1984). For Flanders it is necessary to combine it with the Dutch classification system of van der Werf (1991), 33 forest types defined as Potential Natural Vegetation, especially for coastal dunes, pine and alluvial forest types. Both systems deal with highly developed forest sites. For Flanders a uniform system is being developed. It will deal with all sites (old as well as young forest sites) based on the herbal vegetation of 1500 plots.

Netherlands

Hierarchical (French-Swiss) system of potential natural vegetation types based on German studies in Westphalia, elaborated and extended for The Netherlands by van der Werf (1991). Ground-layer vegetation, rejuvenation and soil type are decisive. Five major types (classes) and 29 forest types (associations).

Denmark

In Denmark, there is no strong tradition for strict plant community systems as in central Europe. But since the middle of the 19th and the beginning of the 20th century forest types based on dominant tree species, soil conditions (mull-mor) have been described in works of e.g. Vaupell, Warming, Bornebusch, Olsen and Raunkiær. In practice in the later years, by different authors a maximum of 10–20 different types have been used including artificial types as plantations of different conifers and types defined by a certain management regime like coppice woods and pasture woods. In 1999 a database of Danish vegetation types, DANVEG (Nygaard et al. 1999), based on computerized statistic classification of a large amount of floristic data (published as well as unpublished flora lists) was elaborated by the National Environmental Research Institute. The database includes 33 types of forest vegetation. The types relates to both the CORINE biotopes, the EU habitat types and the Vegetation Types in the Nordic Countries (Påhlsson 1994).

France

The classification of forests in France is mainly based on a phyto-sociological analysis (syntaxons). The scheme is presented in Rameau (1997). 389 forest types are described (135 broad-leaved forests, 158 conifer forests, 73 flood plain forests and 23 broad-leaved forests).

Germany

There are two main forest typologies actually in use:

The forest typology presently mostly used by vegetation scientists is based on plantsociological methods as founded by Braun-Blanquet. The basic principle is a characterisation of forest communities by the floristic similarity of the vegetation living in them. Ideally, some key species exist, which are directly and without exception associated with certain plant societies. The problem in forests is that underneath a usually rather dense forest canopy only few plant species and individuals can exist in general. Often, only few individuals can be found in a forest stand, some of them even being ubiquists, which cannot be used to describe the floristic characteristics. Therefore, characteristic sets (= combinations) of species with a wider range of occurence, but with typical combinations are being used for plant sociological classification of forests, when characteristic species cannot be found. The system was elaborated in Germany by e.g. Tüxen (northern Germany) and Oberdorfer (southern Germany) and is updated in Ellenberg (1996) and in Pott (1995). Following always the principle of floral similarity, communities are grouped in a hierarchical way to vegetation associations, alliances, orders, and classes.

The second system is frequently in use for practical purposes e.g. for nature conservation, site mapping, and silvicultural application. Forest types are classified according to physiognomical and ecological features, i.e. according to main tree species, to structural features and to ecological characteristics of stands (Hofmann 1994, Schmidt 1995, Ellenberg 1996, Wilmanns 1998).

Site conditions include chemical (soil humidity, soil acidity, base saturation, content of certain chemical elements, e.g. nitrogen) and geographic, climatic (altitude) and meteorological factors (continentality, temperature, sunlight requirement). Plants occurring typically on a certain forest site in central Europe have been listed and statistically arranged into "ecological groups", each group characterising certain site conditions (Ellenberg et al. 1991, Jahn and Hübner in Anon. 1996). Thus, the species within one such ecological group belong together neither from a taxonomic point of view nor do they need to occur together at a given place. Their context is an ecological one. This forest vegetation system of site classification is presented in Jahn and Hübner in Anon. (1996). In order to characterise site factors they consider mainly the regional aspect (geographic gradients from north to south and from west to east), altitudinal belt (expressing the climatic character of a region), soil acidity (describing the nutrient conditions and base saturation) as well as the water budget.

Switzerland

In their classification of Swiss forest vegetation, Ellenberg and Klötzli (1972) defined 71 communities based on floristic, physiognomic and ecological considerations. Since the publication appeared, the classification system has served as a reference for ecological and silvicultural studies. Keller et al. (1998) tested the original vegetation data for statistical reliability, and incorporated site information and structure. The new constancy tables, based on a reanalysis of the original data, differ (in some cases substantially) from the previous tables. The revision of Ellenberg and Klötzli (1972) clarifies the relationships between the previously proposed units. Because Keller et al. (1998) is based on a well-founded existing classification of Swiss forests (and despite some substantial differences from this classification), it provides a consistent and reliable basis for future scientific studies as well as for practical applications.

Austria

Hufnagl´s forest types (1970) are still used in practical site description and forest planning. The classification is based on dominating ground layer vegetation reflecting nutrient conditions and soil moisture but also considering site degradation caused by silvicultural or agricultural influence. Aichinger´s forest development types (1967) stand for a more dynamic approach, considering successional development of stands additionally to site and vegetation characteristics. Both systems are mainly used by practical foresters as tools for silvicultural decision-making like choice of tree species or silvicultural treatment. The first comprehensive classification of the eastern Alps was worked out by Mayer (1974) using a modification of Braun-Blanquet´s method and is still used for practical applications. Mucina et al. (1993) compiled a synopsis of all known plant communities of Austria which is at the moment the most relevant basis for scientific purposes as well as practical applications in silviculture and landscape planning. In general, the Braun-Blanquet school is the prevailing method of vegetation classification in Austria (Braun-Blanquet 1964).

Hungary

In Hungary there are basically two types of forest classification systems. Firstly, the Braun-Blanquet type phytosociological descriptions. This system has changed several times during the past decades, but all versions are based on floristic similarities, and the description of characteristic patches of mostly climax forest vegetation. Major references are: Soó (1960, 1964, 1966, 1968, 1970, 1973, 1980), Bartha et al. (1995), Borhidi (1996). Secondly, a system of forest types sensu Cajander was also developed. Forest management planning was based on these principles in the late 1950s and early 1960s. Unfortunately, later, this system lost its importance for the practice. The major reference for this system is Majer (1968).

Slovenia

Classification of forests are made according to Braun-Blanquet (Ellenberg 1996).

Spain

The forest classification schemes used in Spain can be classified in two groups.

1) Classifications applied in forestry: National Forest Inventory (Anon. 1990), Forest Map of Spain (Ceballos et al. 1966, Ruiz de la Torre 1990), National Map of Crops and Land-Uses (Anon. 1989). They are based mainly on the tree species (or species group) dominant. More than 50 types of unequal importance may be usually differentiated according to this criterion (including plantations), plus mixed types with/without indication of the relative proportions. These types are usually grouped according to physiognomy (e.g.: conifer forests: pine forests/other conifer forests; broad-leaved forests: sclerophyllous/deciduous/riparian forests, etc) and can be subdivided attending to structure or management practices (e.g.: high forest/coppice forests; afforestations, immature forests, etc). A more elaborated system proposed by Ruiz de la Torre (1990) combines climatic, structural and floristic (dominant and subdominant species) criteria and also includes shrublands and seminatural grasslands, but the total number of units really produced has not been still accounted for.

2) Phytosociological classification: these agree with classifications based on dominant trees at higher syntaxonomic levels, but introduce more subdivisions at the association level, according to the floristic composition of the understorey and to climatic, edaphic and phytogeographic factors. Around 170 associations of wooded forests have been recognised in Spain, half of them corresponding to Mediterranean non-riparian forests. Canary Islands add near 20 associations to this number. The outstanding floristic diversity of Iberian shrublands and scrub allows to recognise > 400 associations. Phytosociological units have been used mainly in actual and potential vegetation mapping. At a national scale only the Map of Vegetation Series of Spain 1:400 000 (Rivas-Martínez 1987) is published and availa-

ble. An updating of this classification is available (Rivas-Martínez et al. 1999). The phytosociological classification was also applied to the cartography of habitat types of the 92/43/CEE Directive (Rivas-Martínez et al. 1994), carried out between 1993 and 1995 at the 1:50 000 scale. These maps of habitat types, including most of Spanish forest types, are not published, but they are available in a digitised version implemented on a GIS, at least for official purposes. A project about the revision of these habitat maps (including habitat types not gathered in the 92/43 Directive) is now beginning, founded by the Office for Nature Conservation. The phytosociological classification was also adopted in the National Plan of Environmental Thematic Mapping, started in 1995 but has now stopped.

Italy

Pignatti forest classification scheme (Pignatti 1998), intended mainly to produce a first synthesis on forest vegetation in Italy. Pignatti's classification uses an approach based on the integration of phytosociology with synecology. Forest types are characterised and described according to: ecological factors (e.g. bioclimatic zones and belts, elevation, exposure, soil); structural and compositional features (e.g. canopy composition, tree layers stratification, indicator species); information on current land use and threats for forest type conservation. Five major vegetation formations and 109 forest types are included.

Greece

Greece geographically belongs to south-eastern Europe (Balkan peninsula). The Balkan Peninsula is separated from the rest of Europe by the rivers Savo and Danube, while to the east from Asia by the Black Sea, Sea of Marmara, Bosporos stena, Dardanelia and Aegean Sea. According to Braun-Blanquet (1952) the SE Europe is located in the cross-road of three large floristic areas of the north hemisphere: Euro-Siberian, Mid-European and Irano-Caspian, whereas the higher peaks of the mountains are covered by flora of a sub-alpine type. Since the southeastern Europe is a part of the European continent, it has a close climatic, physiognomic and phytosociological relation with the continental part. Evidence of this relation is the existence of phytosocieties which have close floristic and ecologicall relations, important for a phytosociological classification. Europe is covered by many distinguished vegetation zones which can be characterised according to Braun-Blanquet as orders. The areas covered by various orders correspond to a vegetation zone, which are further divided into sub-zones (based on alliances). These sub-zones, based on relative associations, are divided into secondary sub-zones. Therefore the vegetation units (orders, alliances and associations), defined floristically, are the base for classification of forest vegetation, and strongly influenced by edaphic and climatic parameters (Braun-Blanquet 1952, Dafis 1976, Athanasiadis 1985, Spanos et al. 1999).

Portugal

Mainland

The National Forest Inventory classifies forest according to the dominance of the various tree species or groups, the development stage and the degree of tree cover, resulting in 62 classes that are generally summarised in the following categories: 1) *Pinus pinaster*, 2) *Pinus pinea*, 3) other conifers, 4) *Quercus suber*, 5) *Quercus rotundifolia*, 6) other oaks, 7) *Eucalyptus* spp., 8) *Castanea sativa* and 9) other broad-leaves. For conservation purposes, mapping of forest habitats used the CORINE landcover classification (SNPRCN 1991) also applied in the NATURA 2000 process. Also recent phytosociological studies based the definition of the biogeographical units of the Portuguese mainland where 33 natural forest associations were recognised (Costa et al. 1998 and Rivas-Martínez et al. 1999).

Azores and Madeira

The forest classification schemes used in these archipelagos are analogous to those used in the Portuguese mainland (Anon. 1998 and SNPRCN 1991), with the exception of the characteristic Macaronesian laurel forest. These particular forest have been subject to different phytosociological studies at various depths of analysis. No single study attempted a comprehensive system for the whole Macaronesian forest. A tendency to distinguish three levels (Thermo, Meso and Supra) according to the elevation and humidity has recently been implemented for the Azores (Dias 1996). For Madeira there are studies aiming to update the early works of Sjögren (1972) and Santos (1990).

Appendix 6: BEAR Technical Report 4.

Indicators for forest biodiversity in Europe: proposal for terms and definitions

User's guide

Objectives

This document was produced at the request of the participants of the European Concerted Action BEAR (Indicators for forest biodiversity in Europe).

Its main purpose is to provide an overview of the current variety of **definitions** and of **terms** applying to **concepts** relevant in the context of BEAR. In a later stage, this document may serve as a basis for BEAR partners to establish their own working **terminology**.

Definition: statement which describes a concept and permits its differentiation from other concepts within a system of concepts.

Term: designation of a defined concept in a special language by a linguistic expression. A term may consist of one or more words or even contain symbols.

Concept: a unit of thought constituted through abstraction on the basis of properties common to a set of objects.

Terminology: set of terms representing the system of concepts of a particular subject field. (Source: ISO 1087: 1990 International standard: Terminology – Vocabulary).

General organisation

Most of the concepts defined here were gathered by the BEAR partners during meetings held in summer and autumn 1998. Others, e.g. "species evenness" and "species richness", were added by the compiler as they are also relevant to BEAR.

This document consists of 117 key terms arranged and numbered in English alphabetical order. **Key terms** are English terms which were selected by the compiler to represent the concepts to be defined, for example "biodiversity".

For most key terms, there are one or several **entry terms** in English, German and/or French. For the purpose of this document, entry terms are defined as alternative designations of concepts, for example "biological diversity" or "Biodiversität", or designations of sub-concepts, for example "alpha (α) diversity".

This implies that entry terms under a given key term are not necessarily exact synonyms, as key terms may refer to quite broad concepts. For example, the German entry terms "Forst" and "Wald" cannot be used synonymously. Obviously, such discrepancies are even more marked *between* languages.

For the sake of simplicity for the user, entry terms such as "biological diversity," "Biodiversität" and "alpha (α) diversity" are all to be found under the key term "biodiversity", rather than being listed separately.

Entry and key terms in all three languages are listed in a single alphabetical index (see below).

Limitations

The approach followed for the production of this document was strictly terminological, i.e. it was concerned with the systematisation and representation of concepts. This implies that e.g. *definitions* are provided for the concept of "forest type", but not a *list* of forest types.

The selection of definitions presented here is necessarily **non-exhaustive and arbitrary**, as these definitions were collected only from printed and on-line documents available to the compiler. However, all definitions found in the available sources were included, i.e. the compiler did not deliberately exclude definitions. Further inclusions of terms and/or definitions in any of the three languages (English, German, French) are possible on request.

For some of the German terms for which definitions had been requested by BEAR partners, i.e. "Wuchsgebiet" (*growth region), no equivalent terms have been identified in original documents in English (translations were excluded). One reason for this may be that these terms refer to concepts specific to German-speaking schools of thought in ecology or silviculture. Suggestions for English equivalents – validated by a written source – and/or written sources for definitions in German are welcome. In the meantime, for the sake of transparency for non-German speaking users, tentative English translations are provided to refer to these concepts. They are marked with an * and should not be used unless properly validated.

Michéle Kaennel Dobbertin (kaennel@wsl.ch), Swiss Federal Research Institute WSL, CH-8903 Birmensdorf, Switzerland.

Index of terms compiled

Defined terms in English, German and French are listed alphabetically in this index.

Bold: key terms
Standard: other entry terms

abundance 1
*****adapted species 2**
Akkumulationsindikatoren
  see bioindication (8)
Akkumulationsmonitoren
  see biomonitoring (9)
aktive Bioindikatoren
  see bioindication (8)
aktive Biomonitoren
  see biomonitoring (9)
alien species
  see alien species (3)
alien species 3
allochthon
  see alien species (3)
allogenetische Sukzession
  see succession (105)
allogenic succession
  see succession (105)
alpha (α) diversity
  see biodiversity (7)
Arten-Arealkunde
  see species-area curve (98)
Auenwald
  see riparian (88)
Aufschlag
  see natural regeneration (68)
authenticity 4
autochthon
  see endemic species (29)
autochtonous species
  see endemic species (29)
autogenetische Sukzession
  see succession (105)
autogenic succession
  see succession (105)
Auwald
  see riparian (88)
baseline data 5
Baum
  see tree (110)
*****befitting species 6**
Bestand
  see stand (99)
Bestandestyp
  see stand type (101)
beta (β) diversity
  see biodiversity (7)

Bewirtschaftungseinheit
  see forest management unit (41)
Biodiversität
  see biodiversity (7)
biodiversité
  see biodiversity (7)
biodiversity 7
biogenic succession
  see succession (105)
bioindication 8
Bioindikatoren
  see bioindication (8)
biological diversity
  see biodiversity (7)
Biomarker
  see bioindication (8)
Biomonitoren
  see biomonitoring (9)
biomonitoring 9
biozönotisches Gleichgewicht
  see equilibrium (31)
centre of diversity 10
character species 11
Charakterart
  see character species (11)
climax 12
community 13
competitive equilibrium
  see equilibrium (31)
condition indicator 14
continuum (de végétation)
  see continuum (15)
continuum 15
criteria
  see criterion (16)
criterion 16
critical species
  see critical species (17)
*****differential species 18**
Differentialart
  see *differential species (18)
disturbance 19
disturbance patch 20
disturbance regime 21
diversité biologique
  see biodiversity (7)
ecological diversity 22
ecological ecosystem integrity
  see ecological integrity (23)
ecological indicator species
  see indicator species (53)
ecological integrity 23
ecological or ecosystem resilience
  see resilience (87)
ecological succession
  see succession (105)

ecosystem 24

ecosystem integrity
  see ecological integrity (23)

ecosystem site type 25

edge effect
  see edge effect (26)

edge effect 26

elasticity
  see resilience (87)

endangered habitat 27

endangered species 28

endemic species 29

environmental resource patch 30

equilibrium 31

equilibrium theory
  see equilibrium (31)

espèce caractéristique
  see character species (11)

espèce clef
  see keystone species (57)

espèce critique
  see critical species (17)

exotic species
  see alien species (3)

extinct species 32

extinction 33

extinction risk 34

extirpation 35

flagship species 36

FMU
  see forest management unit (41)

forest 37

forest ecosystem 38

forest health 39

forest management type 40

forest management unit 41

forest type 42

forêt
  see forest (37)

Forst
  see forest (37)

forstliche Betriebsfläche
  see forest management unit (41)

fragmentation 43

fragmentation des habitats
  see fragmentation (43)

functional diversity 44

fundamental niche
  see niche (70)

gamma (γ) diversity
  see biodiversity (7)

genetic diversity 45

Gesundheitszustand des Waldes
  see forest health (39)

grain 46

grain size of a landscape

  see grain (46)

Grenzlinienwirkung
  see edge effect (26)

*growth region 47

guidelines 48

Habitat
  see habitat (49)

habitat 49

habitat fragmentation
  see fragmentation (43)

Hartholzaue
  see riparian (88)

Hemerobie
  see hemeroby (50)

hemeroby 50

Hemerochor
  see alien species (3)

hémérochore
  see alien species (3)

hemerochore
  see alien species (3)

Hinweis
  see indicator (51)

indicator 51

indicator development 52

indicator species
  see indicator species (53)

indicator species 53

indigenous species
  see endemic species (29)

indigenous species
  see endemic species (29)

Indikator
  see indicator (51)

input parameter
  see parameter (73)

insurance value 54

intégrité écologique
  see ecological integrity (23)

intégrité écologique d'un écosystème
  see ecological integrity (23)

intensity of abundance
  see abundance (1)

inventory 55

inventorying
  see inventory (55)

isolation 56

isolement
  see isolation (56)

Kennart
  see character species (11)

keystone species 57

Klimax
  see climax (12)

landscape 58

landscape structure 59

Landschaft
  see landscape (58)
*man-made 60
management indicator species
  see indicator species (53)
matrix 61
Mehrzwecknutzung
  see multiple use (64)
microsite
  see niche (70)
minimum viable population 62
monitoring 63
multiple use 64
multiple-purpose forestry
  see multiple use (64)
multiple-use forestry
  see multiple use (64)
native forest 65
native species
  see endemic species (29)
natural disturbance regime
  see disturbance regime (21)
natural forest
  see natural forest (66)
natural forest 66
natural landscape 67
natural regeneration 68
naturfern
  see *man-made (60)
naturferner Bestand
  see *man-made (60)
naturfremder Bestand
  see *unnatural (113)
naturnah
  see near-natural (69)
naturnaher Bestand
  see near-natural (69)
Naturverjüngung
  see natural regeneration (68)
Naturwald
  see natural forest (66)
near-natural 69
niche 70
niche space
  see niche (70)
Nische
  see niche (70)
non-equilibrium theory
  see equilibrium (31)
non-native species
  see alien species (3)
norm 71
open forest
  see natural forest (66)
output parameter
  see parameter (73)

ownership 72
parameter
parameter 73
passive Bioindikatoren
  see bioindication (8)
passive Biomonitoren
  see biomonitoring (9)
patch 74
patch dynamics 75
patch turnover 76
paysage
  see landscape (58)
plantation 77
Population
  see population (78)
population
  see population (78)
population 78
population viability analysis
  see population vulnerability analysis (79)
population vulnerability analysis 79
prevalence of abundance
  see abundance (1)
primäre Sukzession
  see succession (105)
primary forest 80
principle 81
private ownership (in)
  see ownership (72)
process parameter
  see parameter (73)
provenance 82
public ownership (in)
  see ownership (72)
qualitative indicator
  see indicator (51)
quantitative indicator
  see indicator (51)
rare species 83
rareté
  see rarity (84)
rarity 84
realized niche
  see niche (70)
regenerated patch 85
regeneration
  see regeneration (86)
regeneration 86
régénération naturelle
  see natural regeneration (68)
remnant patch
  see regenerated patch (85)
resilience 87
riparian 88
riparian area
  see riparian (88)

riparian forest
   see riparian (88)
Riparian Management Area (RMA)
   see riparian (88)
ripicole
   see riparian (88)
ripisylve
   see riparian (88)
secondary forest 89
   see bioindication (8)
sekundäre Sukzession
   see succession (105)
sensitive area 90
sensitive slopes
   see sensitive area (90)
sensitive soils
   see sensitive area (90)
sensitive watershed
   see sensitive area (90)
set of disturbance regimes
   see disturbance regime (21)
site 91
*site adaptedness 92
site class 93
site type 94
species diversity 95
species evenness 96
species richness 97
species-area curve 98
stand 99
stand structure 100
stand type 101
standard 102
Standort
   see site (91)
standortfremde Baumart
   see alien species (3)
standortfremde Baumart
   see alien species (3)
standortheimische Baumart
   see endemic species (29)
standortsfremd
   see alien species (3)
standortsgemäss
   see *adapted species (2)
standortsgemäß
   see *adapted species (2)
standortsgerecht
   see *adapted species (2)
standortsheimisch
   see endemic species (29)
Standortsklasse
   see site type (94)
standortstauglich
   see *befitting species (6)
Standortstolerance

   see *site adaptedness (92)
Standortstyp
   see site type (94)
Standortstypengruppe
   see site type (94)
standortswidrig
   see *unsuitable species (114)
standorttaugliche Baumart
   see *befitting species (6)
standortwidrig
   see *unsuitable species (114)
stressor indicator 103
structural diversity 104
succession
   see succession (105)
succession 105
Sukzession
   see succession (105)
Sukzession
   see succession (105)
sustainable forest management 106
temporal diversity 107
texture 108
texture measures
   see texture (108)
threatened or endangered habitats
   see endangered habitat (27)
threatened or endangered species
   see endangered species (28)
threatened species 109
tree 110
umbrella species 111
unitype species 112
*unnatural 113
unstable equilibrium
   see equilibrium (31)
*unsuitable species 114
Urwald
   see primary forest (80)
Vegetation
   see vegetation (115)
vegetation 115
*vegetation type 116
Vegetationstyp
   see *vegetation type (116)
verifier 117
Verjüngung
   see regeneration (86)
vulnerable species 118
Wald
   see forest (37)
Walddefinition
   see forest (37)
Waldform
   see forest management type (40)
Waldzustand

see forest health (39)
Weichholzaue
   see riparian (88)
Weiser
   see indicator (51)
Wuchsgebiet
   see *growth region (47)
Zeiger
   see indicator (51)
Zeigerpflanzen
   see indicator species (53)

Terms and definitions

1 abundance

528 **intensity of abundance**
The number of individuals per habitable site in a community, (cf. Prevalence of abundance).

Source: Begon M., Harper J.L., Townsend C.R., 1996. Ecology: Individuals, populations and communities. Third edition. Oxford, Blackwell. Pp. 1068.

527 **prevalence of abundance**
The proportion or percentage of habitable sites or areas in which a particular species is present.

Source: Begon M., Harper J.L., Townsend C.R., 1996. Ecology: Individuals, populations and communities. Third edition. Oxford, Blackwell. Pp. 1068.

2 *adapted species[□]

568 **standortsgemäß**
Standortsgemäß ist ein Baum oder Baumbestand, wenn er am Ort des Anbaus befriedigende Wuchsleistungen mit ausreichender Stabilität gegenüber abiotischen und biotischen Schadfaktoren vereint und keine nachteiligen Einflüsse auf den Standort hat.
   □ Unverified key terms are marked with *.

Source: Wald und Boden. 1996. Schriftenreihe der Sächsischen Landesanstalt für Forsten 7/96.

583 **standortsgemäss**
= standortsgerecht.
Auch die Schreibweise ohne Genitiv-S im Wortmitte kommt vor.

Source: P. Brang (WSL Birmensdorf).

588 **standortsgerecht**
Auf einem Standort gedeihend und diesen nicht schädigend.
Auch die Schreibweise ohne Genitiv-S im Wortmitte kommt vor.

Source: P. Brang (WSL Birmensdorf).

3 alien species

216 **alien species**
Species of fish or wildlife, deliberately or accidentally introduced in an ecosystem, which [has] become permanently established; alien species often, but not always, have undesirable effects on native species; called "non-native species" in National Forest Management Act regulations (36 CFS 219).

Source: Marcot B.G., Wisdom M.H., Li H.W., Castillo G.C., 1994. Managing for featured, threatened, endangered, and sensitive species and unique habitats for ecosystem sustainability. General Technical Report PNW 329, Portland, OR, USDA Forest Service, Pacific Northwest Research Station, 39 p.

217 **exotic species**
[Alien species], called "non-native species" in National Forest Management Act regulations (36 CFS 219).

Source: Marcot B.G., Wisdom M.H., Li H.W., Castillo G.C., 1994. Managing for featured, threatened, endangered, and sensitive species and unique habitats for ecosystem sustainability. General Technical Report PNW 329, Portland, OR, USDA Forest Service, Pacific Northwest Research Station, 39 p.

322 **hemerochore**
(...) taxa imported to a certain region due to direct or indirect human action.

Source: Sukopp H., 1972. Wandel von Flora und Vegetation in Mitteleuropa unter dem Einfluss des Menschen. Ber. Ldw. 50: 112–139.

218 **non-native species**
Species of fish or wildlife, deliberately or accidentally introduced in an ecosystem, which [has] become permanently established; alien species often, but not always, have undesirable effects on native species; called "non-native species" in National Forest Management Act regulations (36 CFS 219).

Source: Marcot B.G., Wisdom M.H., Li H.W., Castillo G.C., 1994. Managing for featured, threatened, endangered, and sensitive species and unique habitats for ecosystem sustainability. General Technical Report PNW 329, Portland, OR, USDA Forest Service, Pacific Northwest Research Station, 39 p.

509 allochthon

Von außerhalb eines bestimmten Biotops stammend, biotopfremd (fremdbürtig).

Source: Brünig E., Mayer H., 1980. Waldbauliche Terminologie. Wien, Universität für Bodenkultur. Pp. 207.
http://efern.boku.ac.at/forex/wbterm/

494 Hemerochor

(...) Sippen, die nur infolge direkter oder indirekter Mithilfe des Menschen in ein Gebiet gelangt sind.

Source: Sukopp H., 1972. Wandel von Flora und Vegetation in Mitteleuropa unter dem Einfluss des Menschen. Berichte über Landwirtschaft 50: 112–139.

484 standortfremde Baumart

Baumart, die von Natur aus nicht auf dem Standort wächst.

Source: Schütz J.P., Brang P., Bonfils P., Bucher H.U., 1993. Darstellung der Standortansprüche wichtiger Baumarten im Ökogramm und Gesellschaftsanschluss der Baumarten. In: Schmider P., Küper M., Tschander B., Käser B. (eds), Die Waldstandorte im Kanton Zürich. Vdf, Zürich, pp. 254–258.

502 standortfremde Baumart

Bei der Baumartenwahl jene Baumarten, die im Vergleich zu den standortsheimischen und standortstauglichen Baumarten ausgeschaltet werden müssen, um Produktionsverluste zu vermeiden.

Source: Brünig E., Mayer H., 1980. Waldbauliche Terminologie. Wien, Universität für Bodenkultur. Pp. 207.
http://efern.boku.ac.at/forex/wbterm/

583 standortsfremd

Auf einem Standort nicht von Natur aus vorkommend.
Auch die Schreibweise ohne Genitiv-S im Wortmitte kommt vor.

Source: P. Brang (WSL Birmensdorf).

36 hémérochore

(...) espèces qui n'ont pénétré dans une région que grâce à l'aide directe ou indirecte de l'homme.

Source: Sukopp H., 1972. Wandel von Flora und Vegetation in Mitteleuropa unter dem Einfluss des Menschen. Berichte über Landwirtschaft 50: 112–139.

4 authenticity

465 authenticity

Authenticity, as used here, is a reflection of the extent to which a forest corresponds to a naturally functioning forest in terms of composition and ecology.
It will often, but not always, relate to how closely a secondary forest resembles the original natural forest. However, authenticity is more concerned with present conditions than with a theoretical 'original' forest. (...)
Four components are important:
– the composition of tree species and other plant and animal species;
– the pattern of intraspecific variation, as shown in trees by canopy and stand structure, age-class, understory, etc;
– the functioning of plant and animal species in the forest;
– the process by which the forest changes and regenerates itself over time, as demonstrated by disturbance patterns, forest succession, etc.
Authenticity is one of the four general forest criteria identified by WWF.

Source: Dudley N., 1996. Authenticity as a means of measuring forest quality. Biodiversity Letters 3(1): 6–9.

5 baseline data

325 baseline data

Fundamental units of basic inventory information that are crucial for biodiversity conservation planning and management. These are both biotic and abiotic and usually include: (1) the presence and/or abundance of species and other units; (2) other dependent biotic data (e.g. plant cover for macroarthropods); (3) the appropriate influential abiotic variables, and (4) human variables.

Source: Heywood V.H., Watson R.T., Baste I. (eds), 1995. Global biodiversity assessment. Cambridge, Cambridge University Press. Pp. 1140.

6 *befitting species

490 standorttaugliche Baumart

Standortheimische bzw. standortfremde Baumart, die bis zu einem gewissen Bestockungsanteil auf einem Standort gedeiht, ohne diesen zu schädigen.

Source: Schütz J.P., Brang P., Bonfils P., Bucher H.U., 1993. Darstellung der Standortansprüche wichtiger Baumarten im Ökogramm und Gesellschaftsanschluss der Baumarten. In: Schmider P., Küper M., Tschander B., Käser B. (eds), Die Waldstandorte im Kanton Zürich. Vdf, Zürich, pp. 254–258.

585 standortstauglich
= standortsgerecht (s. "adapted species")
Auch die Schreibweise ohne Genitiv-S im Wortmitte kommt vor.

Source: P. Brang (WSL Birmensdorf).

7 biodiversity

243 alpha (α) diversity
Alpha (α) diversity is within-area diversity, measured as the number of species occuring within an area of a given size (Huston 1994). It therefore measures the richness of a potentially interactive assemblage of species.

Source: Bisby F.A., Coddington J., Thrope J.P., Smartt J., Hengeveld R., Edwards P.J., Duffield S.J. (lead authors), 1995. Characterization of biodiversity. In: Heywood V.H., Watson R.T., Baste I. (eds), Global biodiversity assessment. Cambridge, Cambridge University Press, pp. 21–106.

246 alpha (α) diversity
The diversity of species and the complexity of community structure in a particular forest stand or local ecosystem. Species diversity may simply be species richness (number of species) or one of several indices that combine species richness with some measure of relative commonness or rareness (species evenness), or some other measure of the relative abundance of species.

Source: Kimmins J.P., 1997. Forest ecology. 2nd ed. Upper Saddle River, NJ, Prentice Hall. Pp. 596.

244 beta (β) diversity
Beta (β) diversity was introduced by Whittaker (1960) to designate the degree of species change along a given habitat or physiographic gradient. As such it is a measure of between-area diversity. It cannot be expressed in numbers of species because it is a rate or proportion: it is normally represented in terms of the similarity index or of a species turnover rate.

Source: Bisby F.A., Coddington J., Thrope J.P., Smartt J., Hengeveld R., Edwards P.J., Duffield S.J. (lead authors), 1995. Characterization of biodiversity. In: Heywood V.H., Watson R.T., Baste I. (eds), Global biodiversity assessment. Cambridge, Cambridge University Press, pp. 21–106.

247 beta (β) diversity
Variation in species richness or some other measure of alpha species diversity, and of stand-level structural diversity, across a local environmental gradient such as elevation, or soil moisture and fertility gradients. The variation in meas-ures of alpha diversity between the different ecosystem types or seral stages in a local landscape.

Source: Kimmins J.P., 1997. Forest ecology. 2nd ed. Upper Saddle River, NJ, Prentice Hall. Pp. 596.

245 gamma (γ) diversity
Gamma (γ) diversity is also a measure of within-area diversity; however, it usually refers to overall diversity within a large region (Cornell 1985) and its comprehension has direct connotations with dealing with biodiversity at the landscape level.

Source: Bisby F.A., Coddington J., Thrope J.P., Smartt J., Hengeveld R., Edwards P.J., Duffield S.J. (lead authors), 1995. Characterization of biodiversity. In: Heywood V.H., Watson R.T., Baste I. (eds), Global biodiversity assessment. Cambridge, Cambridge University Press, pp. 21–106.

234 biodiversity
Biodiversity is the totality of genes, species, and ecosystems in a region. (...) Biodiversity can be divided into three hierarchical categories – genes, species, and ecosystems– that describe quite different aspects of living systems and that scientists measure in different ways: genetic diversity (...), species diversity (...), ecosystem diversity (...).

Besides ecosystem diversity, many other expressions of biodiversity can be important. These include the relative abundance of species, the age structure of populations, the pattern of communities in a region, changes in community composition and structure over time, and even such ecological processes as predation, parasitism, and mutualism. More generally, to meet specific management or policy goals, it is often important to examine not only compositional diversity – genes, species, and ecosystems– but also diversity in ecosystem structure and function.

Source: World Resources Institute (WRI), The World Conservation Union (IUCN), United Nations Environment Programme (UNEP), 1992. Global biodiversity strategy: guidelines for action to save, study, and use earth's biotic wealth sustainably and equitably. Washington, DC, World Resources Institute (WRI). Pp. 244.

235 biodiversity
Biodiversity is the total variety of life on earth. It includes all genes, species and ecosystems and the ecological processes of which they are part.

Source: Bibby C.J., Collar N.J., Crosby M.J., Heath M.F., Imboden C., Johnson T.H., Long A.J., Stattersfield A.J., Thirgood S.J., 1992. Putting biodiversity on the map: priority areas for global conservation. (2nd printing). Girton, International Council for Bird Preservation. Pp. 90.

236 biodiversity

Biodiversity is the property of living systems of being distinct, that is, different, unlike. The word is a contraction of Biological Diversity, i.e. the diversity of living beings. (...) Diversity is therefore not an entity, a resource, but a property, a characteristic of nature. Species, populations, certain kinds of tissues are resources, but not their diversity as such. (...)

Source: Solbrig O.T., 1994. Biodiversity: an introduction. In: Solbrig O.T., van Emden H.M., van Oordt P.G.W.H.J. (eds), Biodiversity and global change. Wallingford, CAB International, International Union of Biological Sciences, pp. 13–20.

237 biodiversity

The variety and variability of living organisms and the ecological complexes in which they occur; the variety of the world's species, including their genetic diversity and the assemblages they form.

Source: Reid W.V., Miller K.R., 1989. Keeping options alive: the scientific basis for conserving biodiversity. Washington, World Resources Institute. Pp. 128.

238 biodiversity

The variety of living organisms considered at all levels, from genetics through species, to higher taxonomic levels, and including the variety of habitats and ecosystems.

Source: Meffe G.K., Carroll C.R. (eds), 1994. Principles of conservation biology. Sunderland, Ma, Sinauer. Pp. 600.

239 biodiversity

(...) the structural and functional variety of life forms at genetic, population, species, community, and ecosystem levels.

Source: Sandlund O.T., Hindar K., Brown A. H. D. (eds), 1992. Conservation of biodiversity for sustainable development. Oslo, Scandinavian University Press. Pp. 324.

240 biodiversity

Thus we have several kinds of phenomena masquerading under the general rubric of biodiversity. There are (1) ecological diversity: the number of different sorts of organisms present in a local ecosystem; (2) genealogical diversity: the number of taxa within a monophyletic clade – for example, the number of species within a family; (3) phenotypic diversity, or the amount of variation (or differentiation) within or among populations, or within species or still larger taxa. Gould (1989) calls this latter category of diversity "disparity".

Source: Eldredge N., 1992. Where the twain meet: causal intersections between the genealogical and ecological

realms. In: Eldredge N. (ed), Systematics, ecology, and the biodiversity crisis. New York, Columbia University Press, pp. 1–14.

241 biodiversity

The variety, distribution, and abundance of different plants, animals and microorganisms, the ecological functions and processes they perform, and the genetic diversity they contain at local, regional or landscape levels of analysis. Biodiversity has five principal components: (1) genetic diversity (the genetic complement of all living things); (2) taxonomic diversity (the variety of organisms) ; (3) ecosystem diversity (the three dimensional structures on the earth's surface, including the organisms themsevles); (4) functions or ecological services (what organisms and ecosystems do for each other, their immediate surroundings, and for the ecosphere as a whole (i.e. processes and connectedness throught time and space); and (5) the abiotic matrix within which the above exists (the unity of the soil, water, air, and organisms, with each being interdependent on the continued existence of the other).

Source: Dunster K., Dunster J., 1996. Dictionary of natural resource management. Vancouver, UBC Press. Pp. 363.

242 biodiversity

(...) the simplest operational definition of biodiversity can be formulated as the ensemble and the interactions of the genetic, the species and the ecological diversity, in a given place and at a given time (...). It should be stressed that these interactions are of a hierarchical nature.

(...) In summary, hierarchy is a central phenomenon of biodiversity, and there needs to be a general theory integrating the hierarchical levels of biodiversity, how they come to be and interact. In this way biodiversity, rather than amalgamating disconnected pieces of scientific research, will become a transdisciplinary scientific field in its own, the unique trilogy of biodiversity.

Source: di Castri F., Younès T., 1995. Introduction: Biodiversity, the emergence of a new scientific field – its perspectives and constraints. In: di Castri F., Younès T. (eds), Biodiversity, science and development: towards a new partnership. Wallingford, CAB International, pp. 1–11.

326 biodiversity

The term biodiversity refers to the totality of species, populations, communities and ecosystems, both wild and domesticated, that constitute the life of any one area or of the entire planet. It is a more impressive, all-encompassing term than "nature", but it conveys a similar meaning even though it specifically includes cultural modifications of the natural world.

Source: Dasmann R.F., 1991. The importance of cultural

and biological diversity. In: Oldfield M.L., Alcorn J.B. (eds), Biodiversity: culture, conservation, and ecodevelopment. Boulder, Westview Press, pp. 7–15.

327 biodiversity
... biodiversity means the variability among living organisms from all sources and the ecological systems of which they are a part; this includes diversity within species, between species and of ecosystems. Were life to occur on other planets, or living organisms to be rescued from fossils preserved millions of years ago, the concept could include these as well. It can be partitioned, so that we can talk of the biodiversity of a country, of an area, or of an ecosystem, of a group of organisms, or within a single species.

Biodiversity can be set in a time frame so that species extinctions, the disappearance of ecological associations, or the loss of genetic variants in an extant species can all be classed as losses of biodiversity. New elements of life – by mutation, by natural or artificial selection, by speciation or artificial breeding, by biotechnology, or by ecological manipulation – can similarly be viewed as additions to biodiversity.

Source: Bisby F.A., Coddington J., Thrope J.P., Smartt J., Hengeveld R., Edwards P.J., Duffield S.J. (lead authors), 1995. Characterization of biodiversity. In: Heywood V.H., Watson R.T., Baste I. (eds), Global biodiversity assessment. Cambridge, Cambridge University Press, pp. 21–106.

328 biodiversity
Strictly speaking the word biodiversity refers to the quality, range or extent of differences between the biological entities in a given set. In total it would thus be the diversity of all life and is a characteristic or property of nature, not an entity or a resource. But the word has also come to be used in a looser fashion for the set of diverse organisms themselves, i.e. not the diversity of all life on earth, but all life itself. (...)

For the purposes of the Global Biodiversity Assessment, biodiversity is defined as the total diversity and variability of living things and of the systems of which they are a part. This covers the total range of variation in and variability among systems and organisms, at the bioregional, landscape, ecosystem and habitat levels, at the various organismal levels down to species, population and genes. It also covers the complex sets of structural and functionali relationships within and between these different levels of organization, including human action, and their origins and evolution in space and time.

Source: Heywood V.H., Baste I., Gardner K.A. (lead authors), 1995. Introduction. In: Heywood V.H., Watson R.T., Baste I. (eds), Global biodiversity assessment. Cambridge, Cambridge University Press, pp. 1–19.

329 biodiversity
Although there is no point in dreaming of general agreement on, or even rules for, biodiversity terminology, conceptual rigor is necessary within specific contexts.

Source: Haila V., Kouki J., 1994. The phenomenon of biodiversity in conservation biology. Annales Zoologici Fennici 31(1): 5–18.

330 biodiversity
Diversity of the biotic components of ecosystems at the levels of organization, such as genes, species, populations, communities (e.g. tree community or forest ecoytem) and regions (landscape ecosystems, biogeographic units) (for details see Solbrig, 1991a,b) (compare 'ecosystem'). Biodiversity is more than, but includes, species richness. It denotes the entirety of the life-forms in a system at any level. At a-level in forestry it denotes, as dominance diversity, the pattern of mixture of species within a community (evenness or unevenness of mixture) in terms of their contribution to the number of individuals or biomass of the community: at b-level it denotes the patterns of between-community diversity within a geographic unit at the scale of landscapes; at g-level it denotes differences at larger regional scale. Biodiversity at a-level and b-levels are crucially important elements of sustainable forest conservation and management. Biodiversity determines structural diversity and organizational complexity, and is the key to ecological and economic self-sustainability and sustainability of forest management.

Source: Bruenig E.F., 1996. Conservation and management of tropical rainforests: an integrated approach to sustainability. Wallingford, CAB International. Pp. 339.

331 biodiversity
In its broader definition, biodiversity is the diversity of life in all its forms and all its levels of organization (Hunter 1990), including the ecological structures, functions, and processes at all of these levels (Society of American Foresters 1991). In an attempt to provide a more operational definition, Crow et al. (1994) have identified three broad types or subgroups of biodiversity: compositional, structural, and functional. "Compositional diversity" is the variety of items within an area, such as species in a forest stand. "Structural diversity" can be characterized by the vertical or horizontal distribution of plants, plant sizes, or age distributions. "Functional diversity" is characterized by ecological processes, such as nutrient cycling, decomposition, energy flow, and trophic-level relationships. In addition to its different types, biodiversity can also be considered at various hierarchical levels of biological organization. For example, compositional diversity can be viewed at the genetic, species, or ecosystem levels (Probst and Crow 1991).

Source: Roberts M.R., Gilliam F.S., 1995. Patterns and mechanisms of plant diversity in forested ecosystems: implications for forest management. Ecological Applications 5(4): 969–977.

250 biodiversity
Variation in the biotic community. Used synonymously with the term biological diversity. There are many measures of biodiversity: genetic diversity, local species richness and evenness, and local diversity in community structure (alpha diversity); variation in species richness and community structure across the local landscape (beta diversity); and changes over time in these measures of biodiversity (temporal diversity). All of these measures can occur within one landscape unit. Landscape (physical or ecological) diversity provides a framework for regional biodiversity (gamma diversity).

Some people would prefer the term to be restricted to the biological diversity of ecosystems unaltered by human activity, but the term is now so widely used it is unlikely that this will happen.

Source: Kimmins J.P., 1997. Forest ecology. 2nd ed. Upper Saddle River, NJ, Prentice Hall. Pp. 596.

232 biological diversity
Biological diversity (or biodiversity, as it has come to be called) refers to the variety and variability among living organisms and the ecological complexes in which they occur. Diversity can be defined as the number of different items and their relative frequency. For biological diversity, these items are organized at many levels, ranging from chemical structures that are the molecular basis of heredity to complete ecosystems. Thus, the term encompasses different genes, species, ecosystems, and their relative abundance (OTA, 1987).

Source: Panel of the Board on Science and Technology for International Development, 1992. Conserving biodiversity: a research agenda for development agencies. Report of a Panel of the Board on Science and Technology for International Development U.S. National Research Council. (2nd printing). Washington, D.C., National Academy Press. Pp. 127.

253 biological diversity
The diversity of life: diversity in species, behavior, physiology, morphology, life cycles, ecological characteristics, ecological function, habitat requirements, etc.; often used interchangably with biodiversity.

Source: Kimmins J.P., 1997. Forest ecology. 2nd ed. Upper Saddle River, NJ, Prentice Hall. Pp. 596.

332 biological diversity
For the purposes of this Convention:

"Biological diversity" means the variability among living organisms from all sources including, inter alia, terrestrial, marine and other aquatic ecosystems and the ecological complexes of which they are part; this includes diversity within species, between species and of ecosystems.

Source: United Nations Environment Programme (UNEP), 1992. Convention on Biological Diversity. Text and annexes. Geneva, UNEP. Pp. 34.

576 Biodiversität
Die häufigste Verwendung umfasst die drei Aspekte genetische Vielfalt, Vielfalt der Arten und der Ökosysteme. Der Begriff steht damit praktisch gleichbedeutend für die Gesamtheit der belebten Natur.

Source: Brassel P., Brändli U.B. (Red.), (in press). Schweizerisches Landesforstinventar. Ergebnisse der Zweitaufnahme 1993–1995. Bern, Haupt.

38 biodiversité
Précisons que biodiversité est synonyme de diversité biologique. Sous cette notion très globale, on entend la diversité que présente le monde vivant à tous les niveaux :
– la diversité écologique ou diversité des écosystèmes ;
– la diversité spécifique ou diversité interspécifique ;
– la diversité génétique ou diversité intraspécifique.

Ces distinctions ont l'avantage de la commodité, mais il faut se garder de les considérer comme absolues. La biologie moderne tend à effacer les différences entre diversités spécifique et génétique. Et surtout, tous ces niveaux entretiennent des relations complexes, ce qui justifie l'emploi d'un mot nouveau pour désigner l'ensemble.

Source: Chauvet M., Olivier L., 1993. La biodiversité, enjeu planétaire: préserver notre patrimoine génétique. Paris, Sang de la Terre. Pp. 413.

49 diversité biologique
Concept traduit par un indice (il en existe plusieurs), destiné à évaluer, en un lieu donné, la richesse relative en espèces animales et végétales.

Source: Delpech R., Dume G., Galmiche P., Timbal J., 1985. Vocabulaire: typologie des stations forestières. Paris, Institut pour le Développement Forestier. Pp. 243.

8 bioindication

475 Akkumulationsindikatoren
Organismen, die ein bzw. mehrere Elemente und/oder Verbindungen aus ihrer Umwelt anreichern. Effekt- oder Wirkungsindikatoren [sind] Organismen, die auf eine Exposition mit einem bestimmten Element, einer Verbindung oder einer Anzahl von Stoffen mit spezifischen oder

unspezifischen Effekten antworten. Diese Effekte können Änderungen ihrer morphologischen, histologischen oder zellulären Struktur, ihrer stoffwechselphysiologisch-biochemischen Abläufe, ihrer Verhaltens oder ihrer Populationsstruktur umfassen.

Source: Markert B., Oehlmann J., Roth M., 1997. Biomonitoring von Schwermetallen – eine kritische Bestandsaufnahme. Zeitschrift für Ökologie und Naturschutz 6: 1–8.

476 aktive Bioindikatoren
Organismen, die meist aus Laborzuchten stammen und nach Exposition im Untersuchungsgebiet über einen definierten Zeitraum auf Akkumulation, spezifische oder unspezifische Effekte untersucht werden.

Source: Markert B., Oehlmann J., Roth M., 1997. Biomonitoring von Schwermetallen – eine kritische Bestandsaufnahme. Zeitschrift für Ökologie und Naturschutz 6: 1–8.

527 Bioindikator
Pflanzen und Tiere, die auf bestimmte Umwelteinflüsse besonders empfindlich reagieren. So kann z.B. der Rückgang einzelner Pflanzenarten (Flechten) oder Tierarten (Regenwürmer) Hinweise auf eine verstärkte Luftbelastung mit SO_2, Fluorverbindungen, Flugasche oder Metallstäube geben.

Source: Brünig E., Mayer H., 1980. Waldbauliche Terminologie. Wien, Universität für Bodenkultur. Pp. 207.
http://efern.boku.ac.at/forex/wbterm/

577 Bioindikatoren
a) Arten, aus deren Vorkommen oder Eigenschaften auf andere Eigenschaften des Ökosystems geschlossen wird (TWF). Oft werden als Indikatoren bedrohte Arten (z.B. Auerwild) ausgewählt.

b) Pflanzen und Tiere, die auf bestimmte Umwelteinflüsse besonders empfindlich reagieren (WT); vgl. Indikatoren.

Source: Brassel P., Brändli U.B. (Red.), 1999. Schweizerisches Landesforstinventar. Ergebnisse der Zweitaufnahme 1993–1995. Birmensdorf, WSL; Bern, BUWAL; Bern, Haupt. Pp 442.

473 Bioindikatoren
Organismen oder Organismengemeinschaften, deren Gehalte an bestimmten Elementen bzw. Verbindungen und/oder ihre morphologische, histologische oder zelluläre Struktur, ihre stoffwechselphysiologisch-biochemischen Abläufe, ihr Verhalten oder ihre Populationsstruktur(en) sowie Veränderungen dieser Pa-

rameter Informationen über die Quantität der Umwelt(veränderungen) ergeben.

Source: Markert B., Oehlmann J., Roth M., 1997. Biomonitoring von Schwermetallen – eine kritische Bestandsaufnahme. Zeitschrift für Ökologie und Naturschutz 6: 1–8.

480 Biomarker
Meßbarer biologischer Parameter auf suborganismischer (genetischer, enzymatischer, physiologischer, morphologischer) Ebene, dessen strukturelle oder funktionelle Veränderung dazu geeignet ist, Umwelteinflüsse im allgemeinen und Schadstoffeinwirkungen im besonderen qualitativ und z.T. auch quantitativ anzuzeigen.

Beispiele: Enzym- oder Substrainduktion von Cytochrom P-450 und anderer Phase-I-Enzyme durch diverse halogenierte Kohlenwasserstoffe; Inzidenz industriemelanitischer Formen als Marker für Lufverschmutzung; Hautbräunung beim Menschen durch UV-Strahlung; Veränderungen in der morphologischen, histologischen oder Ultrastruktur [sic] von Organismen oder Monitororganen (z.B. Leber, Thymus, Hoden) nach Schadstoffexposition.

Source: Markert B., Oehlmann J., Roth M., 1997. Biomonitoring von Schwermetallen – eine kritische Bestandsaufnahme. Zeitschrift für Ökologie und Naturschutz 6: 1–8.

478 passive Bioindikatoren
Organismen, die aus ihrem natürlichen Lebensraum entnommen und auf Akkumulation, spezifische oder unspezifische Effekte analysiert werden.

Source: Markert B., Oehlmann J., Roth M., 1997. Biomonitoring von Schwermetallen – eine kritische Bestandsaufnahme. Zeitschrift für Ökologie und Naturschutz 6: 1–8.

9 biomonitoring

474 Akkumulationsmonitoren
Organismen, die ein bzw. mehrere Elemente und/oder Verbindungen aus ihrer Umwelt anreichern. Effekt- oder Wirkungsmonitoren [sind] Organismen, die auf eine Exposition mit einem bestimmten Element, einer Verbindung oder einer Anzahl von Stoffen mit spezifischen oder unspezifischen Effekten antworten. Diese Effekte können Änderungen ihrer morphologischen, histologischen oder zellulären Struktur, ihrer stoffwechselphysiologisch-biochemischen Abläufe, ihrer Verhaltens oder ihrer Populationsstruktur umfassen.

Source: Markert B., Oehlmann J., Roth M., 1997. Biomonitoring von Schwermetallen – eine kritische

Bestandsaufnahme. Zeitschrift für Ökologie und Naturschutz 6: 1–8.

477 aktive Biomonitoren
Organismen, die meist aus Laborzuchten stammen und nach Exposition im Untersuchungsgebiet über einen definierten Zeitraum auf Akkumulation, spezifische oder unspezifische Effekte untersucht werden.

Source: Markert B., Oehlmann J., Roth M., 1997. Biomonitoring von Schwermetallen – eine kritische Bestandsaufnahme. Zeitschrift für Ökologie und Naturschutz 6: 1–8.

472 Biomonitoren
Organismen oder Organismengemeinschaften, deren Gehalte an bestimmten Elementen oder Verbindungen und/oder ihre morphologische, histologische oder zelluläre Struktur, ihre stoffwechselphysiologisch-biochemischen Abläufe, ihr Verhalten oder ihre Populationsstruktur(en) sowie Veränderungen dieser Parameter Informationen über die Quantität der Qualität [sic] der Umwelt(veränderungen) ergeben.

Source: Markert B., Oehlmann J., Roth M., 1997. Biomonitoring von Schwermetallen – eine kritische Bestandsaufnahme. Zeitschrift für Ökologie und Naturschutz 6: 1–8.

479 passive Biomonitoren
Organismen, die aus ihrem natürlichen Lebensraum entnommen und auf Akkumulation, spezifische oder unspezifische Effekte analysiert werden.

Source: Markert B., Oehlmann J., Roth M., 1997. Biomonitoring von Schwermetallen – eine kritische Bestandsaufnahme. Zeitschrift für Ökologie und Naturschutz 6: 1–8.

10 centre of diversity

333 centre of diversity
An area with a high number of species, which might be recognized on a global, regional or local scale.

Source: Heywood V.H., Watson R.T., Baste I. (eds), 1995. Global biodiversity assessment. Cambridge, Cambridge University Press. Pp. 1140.

11 character species

192 character species
Indicator of qualitatively defined community type.

Source: Colinvaux P.A., 1993. Ecology 2. 2nd ed. New York, Wiley. Pp. 688.

261 character species
A plant species that is largely restricted to, and is indicative of, a particular plant association or vegetation or ecosystem unit under consideration.

Source: Kimmins J.P., 1997. Forest ecology. 2nd ed. Upper Saddle River, NJ, Prentice Hall. Pp. 596.

469 Charakterart
Nach allen Richtungen differenzierende Arten sind die sog. Kennarten oder Charakterarten (Braun-Blanquet 1951). Diese Charakter- oder Kennarten sind in 3 Stufen zu unterteilen:
• treu: ausschließlich oder nahezu ausschließlich an eine Gesellschaft gebunden;
• fest: Arten mit deutlicher Bindung, eine bestimmte Gesellschaft bevorzugend, aber auch in anderen Gesellschaften, wenn auch mehr oder weniger spärlich, vorhanden;
• hold: in mehreren Gesellschaften ± reichlich vertreten, aber unter Bevorzugung einer bestimmten Gesellschaft.

Source: Winkler S., 1980. Einführung in die Pflanzenökologie. Stuttgart, Gustav Fischer Verlag. Pp. 256.

468 Charakterart
(...) Arte[n], die nur auf diese eine Einheit beschränkt [ist] und in allen anderen fehlen oder schlechter [gedeiht].

Source: Ellenberg H., Klötzli F., 1972. Waldgesellschaften und Waldstandorte der Schweiz. Mitteilungen EAFV 48: 587–930.

531 Charakterart
Pflanzensoziologischer und tierökologischer Begriff für Arten, die in einem größeren Gebiet ganz oder vorzugsweise in einer bestimmten Pflanzenassoziation oder einem bestimmten Biotoptyp vorkommen und in anderen Gesellschaften weitgehend fehlen.

Source: Brünig E., Mayer H., 1980. Waldbauliche Terminologie. Wien, Universität für Bodenkultur. Pp. 207.
http://efern.boku.ac.at/forex/wbterm/

470 Kennart
Nach allen Richtungen differenzierende Arten sind die sog. Kennarten oder Charakterarten (Braun-Blanquet 1951). Diese Charakter- oder Kennarten sind in 3 Stufen zu unterteilen:
• treu: ausschließlich oder nahezu ausschließlich an eine

Gesellschaft gebunden;
• fest: Arten mit deutlicher Bindung, eine bestimmte Gesellschaft bevorzugend, aber auch in anderen Gesellschaften, wenn auch mehr oder weniger spärlich, vorhanden;
• hold: in mehreren Gesellschaften ± reichlich vertreten, aber unter Bevorzugung einer bestimmten Gesellschaft.

Source: Winkler S., 1980. Einführung in die Pflanzenökologie. Stuttgart, Gustav Fischer Verlag. Pp. 256.

35 espèce caractéristique
(…) espèce qui [n'est] propre qu'à [une] seule unité et manque – ou réussit plus mal – chez toutes les autres.

Source: Ellenberg H., Klötzli F., 1972. Waldgesellschaften und Waldstandorte der Schweiz. Mitteilungen EAFV 48: 587–930.

12 climax

262 climax
The end point of autogenic succession: the final seral stage in a sere. A self-replacing community that is relatively stable over several generations of the dominant plant species, or very persistent in comparison with other seral stages. The character of the climax community depends on the frequency and intensity of ecosystem disturbance relative to the rate of autogenic succession for the site. In many humid parts of the world, forests are the climax vegetation. In some northern and high elevation forests, or in very cool humid areas, closed forest is not the climax condition. *Sphagnum* bog (muskeg) or open woodland dominated by ericaceous shrubs is the climax in the very long-term absence of disturbance.

Source: Kimmins J.P., 1997. Forest ecology. 2nd ed. Upper Saddle River, NJ, Prentice Hall. Pp. 596.

541 Klimax
Die oder der Klimax ist eine abstrakte Pflanzengesellschaft, die den gegebenen Umweltbedingungen (Klima, Boden, Topographie, biotische Faktoren, Mensch) optimal angepaßt ist und sich daher dauernd erhält. Die Klimax ist die relativ stabile Optimalphase, das Endstadium der Vegetationsentwicklung, das unter den gegebenen Klimabedingungen erreichbar ist, z.B. Steppe, sommergrüner Laubwald, Eichen-Hainbuchenwald. Die Entwicklung kann eine primäre oder sekundäre Sukzession sein.

Es werden unterschieden: klimatischer, physiognomischer edaphischer und biologischer Klimax. Der Klimax ist produktionsökologisch nicht der Optimalzustand maximaler Produktivität, der meist weit vor dem Klimax-

zustand liegt. Die Klimaxpflanzengesellschaft besitzt aber häufig den höchsten Grad von Diversität innerhalb von Beständen (α-Diversität) und den niedrigsten zwischen Beständen (β-Diversität). Die Bezeichnung Klimax wird meist auf die Assoziation der Schlußgesellschaft begründet, d.h. auf die (bestandesstrukturell) relativ instabile und selten völlig erreichte Endphase (Zerfallsphase) der Sukzessionsentwicklung.

In Wirklichkeit wird durch vielfältige Umwelteinwirkungen und die Einmaligkeit jedes einzelnen Standortes die Entwicklung in unterschiedliche Richtungen gelenkt (Klimaxgruppe statt des hypothetischen Monoklimax) und in der Regel vor Erreichen der Schlußphase wieder rückgängig oder vorübergehend fixiert (arrested).

Source: Brünig E., Mayer H., 1980. Waldbauliche Terminologie. Wien, Universität für Bodenkultur. Pp. 207.
http://efern.boku.ac.at/forex/wbterm/

13 community

264 community
The assemblage of living organisms (plants, animals, microbes) that interact with each other in energy flow and nutrient cycling processes in an ecosystem. The biotic component of a particular ecosystem.

Source: Kimmins J.P., 1997. Forest ecology. 2nd ed. Upper Saddle River, NJ, Prentice Hall. Pp. 596.

429 community
All the organisms that live in a given habitat and affect one another as part of the food web or through their various influences on the physical environment.

Source: Heywood V.H., Watson R.T., Baste I. (eds), 1995. Global biodiversity assessment. Cambridge, Cambridge University Press. Pp. 1140.

14 condition indicator

335 condition indicator
A characteristic of the environment that provides quantitative estimates of the state of ecological resources and is conceptually tied to a value.
(New term 1993; replaces environmental indicator.)

Source: Environmental Monitoring and Assessment Program (EMAP). Master Glossary [online]. Updated 09/15/97 [cited 1998-08-31]. Available from World Wide Web: <http://www.epa.gov/emfjulte/html/glossary.html>

15 continuum

195 continuum
Community unit describing changing species composition along a gradient.

Source: Colinvaux P.A., 1993. Ecology 2. 2nd ed. New York, Wiley. Pp. 688.

202 continuum
Generally refers to a continuous gradient, but specifically when associated with the definite article, it refers to an aspect of vegetation science. A conceptual view of patterns of variability in vegetation wherein there is compositional continuity along environmental gradients. The environmental gradient may not be geographically continuous and may demand piecing together geographically disjunct representatives of intermediate environmental conditions. Also seen as causal of the presumed vegetation continuum is the gradual and continuous process of species invasion and population demise.

Source: Allen T.F.H., Starr T.B., 1982. Hierarchy. Perspectives for ecological complexity. Chicago, University of Chicago Press. Pp. 310.

267 continuum
Variation in species composition along an environmental gradient in which there is no obvious grouping or association of species. Each species is distributed along the gradient solely in response to its genetically controlled tolerances to, and requirements for, factors of the environment.

Source: Kimmins J.P., 1997. Forest ecology. 2nd ed. Upper Saddle River, NJ, Prentice Hall. Pp. 596.

47 continuum (de végétation)
Végétation dont la composition floristique varie d'une manière continue et très progressive, au sein de laquelle il serait impossible de distinguer, sans étude floristico-statistique préalable, des individus d'association.

Source: Delpech R., Dume G., Galmiche P., Timbal J., 1985. Vocabulaire: typologie des stations forestières. Paris, Institut pour le Développement Forestier. Pp. 243.

16 criterion

336 criteria
[Plural] The broad forest values that society seeks to maintain.

Source: Montreal Process Implementation Group, 1998. A framework of regional (sub-national) level criteria and indicators of sustainable forest management in Australia.

Canberra, Commonweath of Australia. Pp. 108.

337 criterion
A criterion is a state or aspect of the dynamic process of the forest ecosystem, or a state of the interacting social system, which should be in place as a result of adherence to a principle of sustainable forest management (or well managed forest). The way criteria are formulated should give rise to a verdict on the degree of compliance in an actual situation.

Source: van Bueren E.M.L., Blom E.M., 1997. Hierarchical framework for the formulation of sustainable forest management standards. Leiden, The Netherlands, Backhuys Publishers. Pp. 82.

338 criterion
A category of conditions or processes by which sustainable forest management may be assessed. A criterion is characterized by a set of related indicators which are monitored periodically to assess change, (Montreal process).

Source: Granholm H., Vähänen T., Sahlberg S. (compilers), 1996. Background document. Intergovernmental Seminar on Criteria and Indicators for Sustainable Forest Management, August 19–22, 1996, Helsinki. Helsinki, Ministry of Agriculture and Forestry, Intergovernmental Seminar on Criteria and Indicators for Sustainable Forest Management. Pp. 131.

339 criterion
Corresponds to an element of sustainability in regard to which forest management can be evaluated, (Tarapoto process).

Source: Granholm H., Vähänen T., Sahlberg S. (compilers), 1996. Background document. Intergovernmental Seminar on Criteria and Indicators for Sustainable Forest Management, August 19–22, 1996, Helsinki. Helsinki, Ministry of Agriculture and Forestry, Intergovernmental Seminar on Criteria and Indicators for Sustainable Forest Management. Pp. 131.

17 critical species

340 critical species
50% probability of extinction within 5 years or 2 generations, whichever is longer. (See "Extinction risk").

Source: Mace G.M., Lande R., 1995. Assessing extinction threats: toward a reevaluation of IUCN threatened species categories. In: Ehrenfeld D. (ed), To preserve biodiversity: an overview. Cambridge, Massachusetts, Blackwell Science, pp. 51–60.

39 espèce critique

Probabilité d'extinction en 5 ans, ou deux générations au maximum, de 50%. (Voir "Extinction risk").

Source: Aird P.L., 1994. Conservation for the sustainable development of forests worldwide – a compendium of concepts and terms. Forestry Chronicle 70(6): 666–674.

18 *differential species

532 Differentialart

Pflanzensoziologischer und tierökologischer Begriff für die sich in ihrem Vorkommen in einem bestimmten Gebiet fast oder ganz ausschließenden (d.h. ökologisch vikariierenden) Arten. Sie dienen zur Unterscheidung und Kennzeichnung nahe verwandter Pflanzengesellschaften bzw. Lebensgemeinschaften.

Source: Brünig E., Mayer H., 1980. Waldbauliche Terminologie. Wien, Universität für Bodenkultur. Pp. 207.
http://efern.boku.ac.at/forex/wbterm/

19 disturbance

206 disturbance

An event that causes a significant change from the normal pattern in an ecological system.

Source: Forman R.T.T., Godron M., 1986. Landscape ecology. New York, Wiley. Pp. 619.

444 disturbance

A discrete event, either natural or human-induced, that causes a change in the existing condition of an ecological system.

Source: British Columbia Forest Service, March 1997: Glossary of Forestry Terms. Available from the World Wide Web:
http://www.for.gov.bc.ca/pab/publctns/glossary/GLOSSARY.HTM. Cited 01-Oct-98.

502 disturbance

In community ecology, an event that removes organisms and opens up space which can be colonized by individuals of the same or different species.

Source: Begon M., Harper J.L., Townsend C.R., 1996. Ecology: Individuals, populations and communities. Third edition. Oxford, Blackwell. Pp. 1068.

521 disturbance

An event that significantly alters the pattern of variation in the structure or function of a system.
Usually this refers to natural phenomena, and "human activity" is used instead of "human disturbance".

Source: Forman R.T.T., 1995. Land mosaics. The ecology of landscapes and regions. Cambridge, Cambridge University Press. Pp. 632.

20 disturbance patch

205 disturbance patch

An area that has been disturbed within a matrix.

Source: Forman R.T.T., Godron M., 1986. Landscape ecology. New York, Wiley. Pp. 619.

21 disturbance regime

488 set of disturbance regimes

The intensities, frequencies, and types of perturbations (disturbances) characterizing each ecosystem type in a cluster of ecosystem types.

Source: Forman R.T.T., Godron M., 1986. Landscape ecology. New York, Wiley. Pp. 619.

522 natural disturbance regime

Natural disturbances fall into two categories: abiotic and biotic. (...) The characteristics of these natural disturbance agents combine to define a natural disturbance regime based on the area affected, the return interval, and the magnitude of the disturbances.

Source: Parminter J., 1998. Natural disturbance ecology. In: J. Voller, S. Harrison (eds), Conservation biology principles for forested landscapes. Vancouver, UBC, pp. 3–41.

22 ecological diversity

272 ecological diversity

Variation in the physical characteristics of ecosystems across a landscape caused by variations in soil, slope, aspect, elevation, climate, and geology, and the accompanying variation in biotic communities. Ecological diversity provides the ecological framework within which biological diversity develops. Much of the biological diversity across landscapes is a direct reflection of ecological diversity.

Source: Kimmins J.P., 1997. Forest ecology. 2nd ed. UpperSaddle River, NJ, Prentice Hall. Pp. 596.

23 ecological integrity

342 ecological integrity
The quality of a natural unmanaged or managed ecosystem in which the natural ecological processes are sustained, with genetic, species and ecosystem diversity assured for the future.

Source: Aird P.L., 1994. Conservation for the sustainable development of forests worldwide – a compendium of concepts and terms. Forestry Chronicle 70(6): 666–674.

525 ecosystem integrity
The maintenance of an ecosystem within the range of conditions or seral stages in which the processes of autogenic succession operate normally to return the ecosystem to, or toward, its predisturbance condition. Ecosystem integrity is very different from the integrity of a particular seral stage or condition, such as the integrity of the old growth condition. An ecosystem that has been regressed from an old growth condition to an earlier seral stage may not have experienced any loss of ecosystem integrity, but there will have been a loss in the integrity of the old growth condition of that ecosystem.

Source: Kimmins J.P., 1997. Forest ecology. 2nd ed. Upper Saddle River, NJ, Prentice Hall. Pp. 596.

37 intégrité écologique
Qualité d'un écosystème naturel aménagé ou non où se produisent des phénomènes écologiques naturels et où la diversité génétique, spécifique et écosystémique est assurée pour l'avenir.

Source: Aird P.L., 1994. Conservation for the sustainable development of forests worldwide – a compendium of concepts and terms. Forestry Chronicle 70(6): 666–674.

40 intégrité écologique d'un écosystème
= intégrité écologique.

Source: Aird P.L., 1994. Conservation for the sustainable development of forests worldwide – a compendium of concepts and terms. Forestry Chronicle 70(6): 666–674.

24 ecosystem

343 ecosystem
The aggregate of all living organisms and their interactions with each other and the non-living parts of the environment for a defined place or kind of habitat.

Source: Montreal Process Implementation Group, 1998. A framework of regional (sub-national) level criteria and indicators of sustainable forest management in Australia. Canberra, Commonweath of Australia. Pp. 108.

344 ecosystem
A dynamic complex of plant, animal, fungal, and micro-organism communities and their associated non-living environment interacting as an ecological unit; the organisms living in a given environment, such as a tropical forest or a lake, and the physical part of the environment that impinges on them.

Source: Heywood V.H., Watson R.T., Baste I. (eds), 1995. Global biodiversity assessment. Cambridge, Cambridge University Press. Pp. 1140.

443 ecosystem
A functional unit consisting of all the living organisms (plants, animals, and microbes) in a given area, and all the non-living physical and chemical factors of their environment, linked together through nutrient cycling and energy flow. An ecosystem can be of any size – a log, pond, field, forest, or the earth's biosphere – but it always functions as a whole unit. Ecosystems are commonly described according to the major type of vegetation, for example, forest ecosystem, old-growth ecosystem, or range ecosystem.

Source: British Columbia Forest Service, March 1997: Glossary of Forestry Terms. Available from the World Wide Web:
http://www.for.gov.bc.ca/pab/publctns/glossary/GLOSSARY.HTM. Cited 01-Oct-98.

524 ecosystem
An ecological system composed of living organisms (plants, animals, and microbes) and their nonliving environment (climate and soil in the case of terrestrial ecosystems; aqueous environment and substrate in aquatic ecosystems). To be an ecosystem, these components must be spatially arranged and have the appropriate interactions that lead to the capture and storage of energy as biomass, a trophic structure, a circulation of nutrients, and change over time (ecological succession). Ecosystems are characterized by five major attributes: structure, function, complexity, interaction of the components, and change over time.

Source: Kimmins J.P., 1997. Forest ecology. 2nd ed. Upper Saddle River, NJ, Prentice Hall. Pp. 596.

25 ecosystem site type

276 ecosystem site type
A specific type of ecosystem, characterized by relatively homogeneous soil conditions, microclimatic conditions, a characteristic climax plant association, and associated animals and microbes. Generally the smallest, most homoge-

neous unit of ecosystem classification. A subdivision of an ecosystem (site) association according to variation in soil condition.

Source: Kimmins J.P., 1997. Forest ecology. 2nd ed. Upper Saddle River, NJ, Prentice Hall. Pp. 596.

26 edge effect

345 edge effect
Processes that characterize habitat fragmentation and the concomitant creation of edges.

Source: Heywood V.H., Watson R.T., Baste I. (eds), 1995. Global biodiversity assessment. Cambridge, Cambridge University Press. Pp. 1140.

536 Grenzlinienwirkung
Meist günstige Wirkung von Grenzbereichen benachbarter-unterschiedlicher Biotoptypen auf die Besiedlungsmöglichkeit für Wildtiere (z.B. Wald- oder Bestandesränder, Wald-Wiesen- und Feld-Brachland-Grenzbereiche). Da die Habitat-Ansprüche vieler Wildtierarten nicht in einem Biotop (-typ) befriedigt werden können, sondern nur in verschiedenen (z.B. Äsung auf Frei- oder Verjüngungsflächen, Deckung in Dickungen, Suhlen in sumpfigem Gelände, usw.), ist insbesondere für standorttreue, revierbildende Wildarten (wie z.B. Rehwild, Auerwild, viele Singvögel) die Tragfähigkeit eines Verbreitungsgebietes um so höher, je vielfältiger die Gemengelage der hauptsächlich von ihnen genutzten Vegetationsgesellschaften bzw. Biotope ist. Wo immer zwei unterschiedliche Biotoptypen aneinandergrenzen, bildet der Grenzlinien oder Randzonenbereich dieser beiden ein insgesamt günstigeres Wildtierhabitat als jeder dieser noch so günstigen Biotope für sich allein genommen.

Source: Brünig E., Mayer H., 1980. Waldbauliche Terminologie. Wien, Universität für Bodenkultur. Pp. 207. http://efern.boku.ac.at/forex/wbterm/

537 Randzonenwirkung

Source: Brünig E., Mayer H., 1980. Waldbauliche Terminologie. Wien, Universität für Bodenkultur. Pp. 207.
http://efern.boku.ac.at/forex/wbterm/

27 endangered habitat

461 threatened or endangered habitats
Ecosystems that are:
– restricted in their distribution over a natural landscape (e.g., freshwater wetlands within certain biogeo-

climatic) or are restricted to a specific geographic area or a particular type of local environment; or
– ecosystems that were previously widespread or common but now occur over a much smaller area due to extensive disturbance or complete destruction by such practices as intensive harvesting or grazing by introduced species, hydro projects, dyking, and agricultural conversion.

Source: British Columbia Forest Service, March 1997: Glossary of Forestry Terms. Available from the World Wide Web:
http://www.for.gov.bc.ca/pab/publctns/glossary/GLOSSARY.HTM. Cited 01-Oct-98.

28 endangered species

346 endangered species
20% probability of extinction within 20 years or 10 generations, whichever is longer.

Source: Mace G.M., Lande R., 1995. Assessing extinction threats: toward a reevaluation of IUCN threatened species categories. In: Ehrenfeld D. (ed), To preserve biodiversity: an overview. Cambridge, Massachusetts, Blackwell Science, pp. 51–60.

347 endangered species
Taxa in danger of extinction and whose survival is unlikely if the causal factors continue operating. Included are taxa whose numbers have been reduced to a critical level or whose habitats have been so drastically reduced that they are deemed to be in immediate danger of extinction. Also included are taxa that may be extinct but have definitely been seen in the wild in the past 50 years.

Source: UCN Conservation Monitoring Centre, Cambridge U.K. (compiler), 1994. IUCN red list of threatened animals. Gland, IUCN.

348 endangered species
Species in danger of extinction and whose survival is unlikely if causal factors continue operating.

Source: Montreal Process Implementation Group, 1998. A framework of regional (sub-national) level criteria and indicators of sustainable forest management in Australia. Canberra, Commonweath of Australia. Pp. 108.

462 threatened or endangered species
Species identified as red listed by the Ministry of Environment, Lands and Parks; these are indigenous species that are either threatened or endangered.

Source: British Columbia Forest Service, March 1997:

Glossary of Forestry Terms. Available from the World Wide Web: http://www.for.gov.bc.ca/pab/publctns/glossary/GLOSSARY.HTM. Cited 01-Oct-98.

29 endemic species

349 endemic species
A species originating in, or belonging to, a particular region.
Both "endemic" and "indigenous" are the adjectives preferred over "native".

Source: Aird P.L., 1994. Conservation for the sustainable development of forests worldwide – a compendium of concepts and terms. Forestry Chronicle 70(6): 666–674.

508 autochthon
Bodenständig, biotopeigen, indigen im selben Gebiet oder Biotop entstanden.

Source: Brünig E., Mayer H., 1980. Waldbauliche Terminologie. Wien, Universität für Bodenkultur. Pp. 207.
http://efern.boku.ac.at/forex/wbterm/

377 indigenous species
A species originating in, or belonging to, a particular region.
Both "endemic" and "indigenous" are the adjectives preferred over "native".

Source: Aird P.L., 1994. Conservation for the sustainable development of forests worldwide – a compendium of concepts and terms. Forestry Chronicle 70(6): 666–674.

378 indigenous species
Species or genotypes which have evolved in the same area, region or biotope and are adapted to the specific ecological conditions predominant at the time of establishment.

Source: Loiskekosti M., Halko L. (eds), 1993. Ministerial Conference on the Protection of Forests in Europe, 16–17 June 1993 in Helsinki. European list of criteria and most suitable quantitative indicators. Helsinki, Ministry of Agriculture and Forestry. Pp. 20.

390 native species
A species originating in, or belonging to, a particular region.
Both "endemic" and "indigenous" are the adjectives preferred over "native".

Source: Aird P.L., 1994. Conservation for the sustainable development of forests worldwide – a compendium of concepts and terms. Forestry Chronicle 70(6): 666–674.

489 standortheimische Baumart
Baumart, die von Natur aus auf einem Standort vorkommt.

Source: Schütz J.P., Brang P., Bonfils P., Bucher H.U., 1993. Darstellung der Standortansprüche wichtiger Baumarten im Ökogramm und Gesellschaftsanschluss der Baumarten. In: Schmider P., Küper M., Tschander B., Käser B. (eds), Die Waldstandorte im Kanton Zürich. Vdf, Zürich, pp. 254–258.

584 standortsheimisch
Auf einem Standort von Natur aus vorkommend.
Auch die Schreibweise ohne Genitiv-S im Wortmitte kommt vor.

Source: P. Brang (WSL Birmensdorf).

30 environmental resource patch

207 environmental resource patch
An area where environmental resources, such as soil moisture or rock type, differ from the surrounding matrix.

Source: Forman R.T.T., Godron M., 1986. Landscape ecology. New York, Wiley. Pp. 619.

31 equilibrium

519 competitive equilibrium
A persistent coexistence of species resulting directly from competition.

Source: Forman R.T.T., Godron M., 1986. Landscape ecology. New York, Wiley. Pp. 619.

431 equilibrium theory
Theory that suggests that under natural circumstances, species addition and loss are balanced, and furthermore, that displacement from the equilibrium value results in changes in speciation or extinction rate that tend to restore the system to its equilibrium rate.

Source: Heywood V.H., Watson R.T., Baste I. (eds), 1995. Global biodiversity assessment. Cambridge, Cambridge University Press. Pp. 1140.

529 biozönotisches Gleichgewicht
Das dynamische Abhängigkeits- und Wirkungsgefüge in einer Lebensgemeinschaft (Biozönose), das trotz der Bevölkerungsschwankungen der einzelnen Arten oder anderer Störfaktoren die Stabilität des Gesamtsystems erhält, solange nicht grundsätzliche Milieuänderungen eintreten; vgl. Biologisches Gleichgewicht.

Source: Brünig E., Mayer H., 1980. Waldbauliche Terminologie. Wien, Universität für Bodenkultur. Pp. 207. http://efern.boku.ac.at/forex/wbterm/

434 non-equilibrium theory
Suggests that the number of species increases or decreases depending on how the environment influences species production, exchange and extinction at any particular time.

Source: Heywood V.H., Watson R.T., Baste I. (eds), 1995. Global biodiversity assessment. Cambridge, Cambridge University Press. Pp. 1140.

513 unstable equilibrium
In an ecological context, a level of a population, or populations, or of resources, from which slight displacements lead to larger displacements.

Source: Begon M., Harper J.L., Townsend C.R., 1996. Ecology: Individuals, populations and communities. Third edition. Oxford, Blackwell. Pp. 1068.

32 extinct species

351 extinct species
Species not definitely located in the wild during the past 50 years.

Source: UCN Conservation Monitoring Centre, Cambridge U.K. (compiler), 1994. IUCN red list of threatened animals. Gland, IUCN.

33 extinction

352 extinction
The termination of a species caused by failure to reproduce and death of all remaining members of the species.

Source: Aird P.L., 1994. Conservation for the sustainable development of forests worldwide – a compendium of concepts and terms. Forestry Chronicle 70(6): 666–674.

353 extinction
The death of any lineages of organisms. Extinction can be local, (when it is known as extirpation) in which one population of a given species vanishes while others survive elsewhere, or total, in which all its populations vanish.

Source: Heywood V.H., Watson R.T., Baste I. (eds), 1995. Global biodiversity assessment. Cambridge, Cambridge University Press. Pp. 1140.

34 extinction risk

354 extinction risk
The risk of premature extinction may be defined in terms of the probability of extinction within a specific time period. It is based on the theory of extinction times for single populations and on meaningful time scales for conservation action. Three categories are defined on the basis of decreasing probabilities of extinction risk over increasing time periods: critical species, endangered species, vulnerable species.

Source: Aird P.L., 1994. Conservation for the sustainable development of forests worldwide – a compendium of concepts and terms. Forestry Chronicle 70(6): 666–674.

35 extirpation

355 extirpation
Local extinction; a species or subspecies disappearing from a locality or region without becoming extinct throughout its range (McNeely et al. 1990).

Source: Aird P.L., 1994. Conservation for the sustainable development of forests worldwide – a compendium of concepts and terms. Forestry Chronicle 70(6): 666–674.

36 flagship species

221 flagship species
A highly charismatic species, typically a large-bodied mammal or bird, that is in some peril of extirpation and that can be managed so as to also provide habitats and resources for other species; compare with umbrella species.

Source: Marcot B.G., Wisdom M.H., Li H.W., Castillo G.C., 1994. Managing for featured, threatened, endangered, and sensitive species and unique habitats for ecosystem sustainability. General Technical Report PNW 329, Portland, OR, USDA Forest Service, Pacific Northwest Research Station, 39 p.

356 flagship species
Popular, charismatic species that serves as symbol and rallying point to stimulate conservation awareness and action.

Source: Heywood V.H., Watson R.T., Baste I. (eds), 1995. Global biodiversity assessment. Cambridge, Cambridge University Press. Pp. 1140.

37 forest

180 forest
1. Ecology: Generally, an ecosystem characterized by a more or less dense and extensive tree cover. More particularly, a plant community predominantly of trees and other woody vegetation, growing more or less closely together.
2. Silviculture/Forest management: An area managed for the production of timber and other forest produce, or maintained under woody vegetation for such indirect benefits as the protection of watersheds, the povision of recreation areas, or the preservation of natural habitat.

Source: Policy and Economics Directorate Forestry Canada / Direction Générale des politiques et de l'Economie, Forêts Canada, 1992. Silvicultural terms in Canada / Terminologie de la sylviculture au Canada (trad. et adapté de la version anglaise). Ottawa, Science and Sustainable Development Directorate, Forestry Canada. Pp. 74.

184 forest
An ecosystem characterized by a more or less dense and extensive tree cover often consisting of stands varying in characteristics such as species composition, structure, age class, and associated processes and commonly including meadows, streams, fish, and wildlife.
  Forests include special kinds such as industrial forests, non-industrial private forests, plantations, public forests, protection forests, urban forests, as well as parks and wilderness.

Source: Helms J.A., 1998. The Dictionary of Forestry. Bethesda, MD, Society of American Foresters. Pp. 224.

185 forest
1. Ecology: An ecosystem characterized by a more or less dense and extensive tree cover.
2. Ecology: A plant community predominantly of trees and other woody vegetation, growing more or less closely together.
3. Silviculture, management: An area managed for the production of timber and other forest produce, or maintained under woody vegetation for such indirect benefits as protection of catchment areas or recreation.
  Forests include special kinds such as industrial forests, non-industrial private forests, plantations, public forests, protection forests, urban forests, as well as parks and wilderness.

Source: Ford-Robertson F.C. (ed), 1983. Terminology of forest science, technology, practice and products. 2nd printing with addendum. Washington, D.C., Society of American Foresters. Pp. 370.

187 forest
A plant community predominantly of trees and other woody vegetation, growing more or less closely together.
  Forests include special kinds such as industrial forests, non-industrial private forests, plantations, public forests, protection forests, urban forests, as well as parks and wilderness.

Source: Haddon B.D. (ed), 1988. Forest inventory terms in Canada / Terminologie de l'inventaire des forêts du Canada. 3rd ed. Ottawa, Canadian Forest Inventory Committee, Forestry Canada. Pp. 113 + 109

191 forest
A complex assemblage of plants, animals, and environment dominated by trees.

Source: Jensen E.C., Anderson D.J., 1995. The reproductive ecology of broadleaved trees and shrubs: glossary. Research Contribution 9f, Corvallis, Forest Research Laboratory, Oregon State University, 8 p.

196 forest
Land which is at least 10% stocked with any combination of trees found in the FHM [Forest Health Monitoring] species and at least one acre in size. Adjacent non-forest inclusions less than one acre in size are considered forest land.

Source: Tallent-Halsell N.G. (ed), 1994. Forest health monitoring 1994. Field methods guide. EPA/620/R-94/027. Washington, D.C., U.S. Environmental Protection Agency. Pp. [mult. pag.].

197 forest
Land which is at least 10% stocked with any combination of trees found in the FHM [Forest Health Monitoring] species and at least one acre (0.4 ha) in size and 120 ft. (36.6 m) wide. Adjacent nonforest inclusions less than one acre in size are considered forest land.

Source: USDA Forest Service, 1998. Forest health monitoring 1998. Field methods guide. Research Triangle Park, NC, USDA Forest Service, National Forest Health Monitoring Program. Pp. [mult. pag.].

226 forest
(...) The legal definitions of what has to be considered as forest land differ between countries (...). The quantitative criteria found in most definitions for forest area are crown cover (10% in France, 30% in Austria, 20% in Switzerland, in Scandinavia not applied), width of the stand (10 to 25 m) and minimum area (0.05 ha ... 0.5 ha). In general, clearcut areas and young stands which do not fulfil the requirements are included in the forest area. (...)

Source: Köhl M., Päivinen R., 1996. Definition of a system of nomenclature for mapping European forests and for compiling a pan-European forest information system. Luxembourg, Office for Official Publications of the European Communities. Pp. 238.

358 forest
Land with tree crown cover (stand density) of more than about 20% of the area. Continuous forest with trees usually growing to more than about 7 m in height and able to produce wood [sic]. This includes both closed forest formations where trees of various layers and undergrowth cover a high proportion of the ground and open forest formations with a continuous grass layer in which tree synusia cover at least 10% of the ground.

Source: Loiskekosti M., Halko L. (eds), 1993. Ministerial Conference on the Protection of Forests in Europe, 16–17 June 1993 in Helsinki. European list of criteria and most suitable quantitative indicators. Helsinki, Ministry of Agriculture and Forestry. Pp. 20.

359 forest
An area, incorporating all living and non-living components, that is dominated by trees having usually a single stem and a mature or potentially mature stand height exceeding two metres and with existing or potential crown cover of overstorey strata about equal to or greater than 20 per cent. This definition includes Australia's diverse native forests and plantations, regardless of age. It is also sufficiently broad to encompass areas of trees that are sometimes described as woodlands.

Source: Montreal Process Implementation Group, 1998. A framework of regional (sub-national) level criteria and indicators of sustainable forest management in Australia. Canberra, Commonweath of Australia. Pp. 108.

445 forest
As defined by the Forest Practices Code of British Columbia Act includes all of the following: forest land, whether Crown land or private land; Crown range; Crown land or private land that is predominantly maintained in one or more successive stands of trees, successive crops of forage, or wilderness.

Source: British Columbia Forest Service, March 1997: Glossary of Forestry Terms. Available from the World Wide Web: http://www.for.gov.bc.ca/pab/publctns/glossary/ GLOSSARY.HTM. Cited 01-Oct-98.

490 forest
(Definition applied for developed countries.) Land with tree crown cover (stand density) of more than about 20 percent of the area. Continous forest with trees usually growing to more than about 7 m in height and able to produce wood. This includes both closed forest formations where trees of various storeys and undergrowth cover a high proportion of the ground, and open forest formations with a continous grass layer in which tree synusia cover at least 10 percent of the ground.

Source: Food and Agriculture Organization (ed), 1997. State of the world's forests 1997. Rome, Food and Agriculture Organization. Pp. 200. http://www.fao.org/waicent/faoinfo/forestry/ SOFOTOC.htm

491 forest
(Definition applied for developing countries.) Ecosystem with a minimum of 10 percent crown cover of trees and/or bamboos, generally associated with wild flora, fauna and natural soil conditions, and not subject to agricultural practices. The term forest is further subdivided, according to its origin, into two categories:
 i) Natural forests: a subset of forests composed of tree species known to be indigenous to the area; and
 ii) Plantation forests:
 – established artificially by afforestation on lands which previously did not carry forest within living memory;
 – established artificially by reforestation of land which carried forest before, and involving the replacement of the indigenous species by a new and essentially different species or genetic variety.

Source: Food and Agriculture Organization (ed), 1997. State of the world's forests 1997. Rome, Food and Agriculture Organization. Pp. 200. http://www.fao.org/waicent/faoinfo/forestry/ SOFOTOC.htm

498 forest
Austria:
Forests are areas of min 1000 m^2 and cover percentage 30% with tree species, according to forest law.

Source: Köhl M., Päivinen R., 1996. Definition of a system of nomenclature for mapping European forests and for compiling a pan-European forest information system. Luxembourg, Office for Official Publications of the European Communities. Pp. 238.

499 forest
Forest in the sense of the National Forest Inventory/Forest Law in Germany is, independently of the entries on cadastral maps or similar inventories, every are stocked with forest plants.

Forests also include clearcut or opened areas, forest tracks and roads, forest meadows, game grazing areas, wood storage areas, corridors for utilities (e.g. pylone), as well as recreation structures connected to the forest, heaths and bogs that are reverting to forests, former meadows and pastures, alpine grazing and other roughgrazing that are reverting to forests and areas of "krummholz" pine and green alder.

Heaths, bogs, meadows, alpine grazing and rough grazing count as reverted to forest when the cover of resprouting trees reaches at least 50% and a mean age of 5 years. Agricultural land and built-up areas with woodland areas under 1000 m², woodland strips under 10 m wide, christmas tree plantations and other areas such as gardens belong to parkland and are not considered as forest by the NFI. Rivers less than 5 m wide do not constitute a division for a forest area.

Source: Bundesministerium für Ernährung, Landwirtschaft und Forsten: Bundesinventur 1986–1990, Vol. I, p. 115. Translation and comments: J. Innes, H. Ellenberg, Jyllinge. Cited in: Frank G., Halbritter K., 1998: Regional BEAR support meeting, 11–12 July 1998, Göttingen, Germany. Minutes. 15 pp.

529 forest
Land with tree crown cover (or equivalent stocking level) of more than 10 percent and area of more than 0.5 ha. The trees should be able to reach a minimum height of 5 m at maturity in situ. May consist either of closed forest formations where trees of various storeys and undergrowth cover a high proportion of the ground; or of open forest formations with a continuous vegetation cover in which tree crown cover exceeds 10 percent. Young natural stands and all plantations established for forestry purposes which have yet to reach a crown density of 10 percent or tree height of 5 m are included under forest, as are areas normally forming part of the forest area which are temporarily unstocked as a result of human intervention or natural causes but which are expected to revert to forest.

Includes: Forest nurseries and seed orchards that constitute an integral part of the forest; forest roads, cleared tracts, firebreaks and other small open areas within the forest; forest in national parks, nature reserves and other protected areas such as those of special environmental, scientific, historical, cultural or spiritual interest; windbreaks and shelterbelts of trees with an area of more than 0.5 ha and a width of more than 20 m. Rubberwood plantations and cork oak stands are included.

Excludes: Land predominantly used for agricultural practices.

Source: UN-ECE/FAO, 1997. UN-ECE/FAO Temperate and Boreal Forest Resources Assessment 2000. Terms and definitions. New York and Geneva, United Nations. Pp. 13.

535 Forst
Im deutschen Sprachgebrauch auf den zur Produktion von Rohstoffen und Infrastrukturleistungen bewirtschafteten Wald bezogen. Engl. "forest" umfaßt dagegen auch den Urwald (z.B. tropical rain forest).

(1) Einforsten (forest reservation, establishment of permanent forests). Die Begründung und Sicherung aufgrund hoheitlicher Rechte (forstgesetzliche Bestimmungen) und flächenmäßige Festlegung und Markierung (demarcation) von Landflächen zum Zweck der forstlichen Bewirtschaftung.

(2) Forstliche Landfläche (forested land, forest land). In der Flächenstatistik alle Flächen, welche von Bäumen beherrscht werden, gleichgültig, ob sie genutzt werden oder nicht, soweit sie in der Lage sind, Holz und Nebenprodukte zu produzieren, einen Einfluß auf Klima und Wasserhaushalt auszuüben und Schutz für Vieh und Wild bieten zu können. Durch Kahlhieb baumfrei gewordene Flächen sind eingeschlossen, wenn diese in absehbarer Zeit wieder bestockt werden. Außerdem sind eingeschlossen Savannen bis zu einem Beschirmungsgrad von mindestens 0,05 und Flächen, die nach Wanderfeldbau liegengelassen wurden und sich in einer Rückentwicklung zur ursprünglichen Waldvegetation befinden (sekundäre Sukzession), sowie andere degradierte Vegetationsformen wie z.B. Gebüsche und Gehölze auf Standorten, die ohne den Einfluß der Degradationsfaktoren mit Wald bestockt sein würden.

Source: Brünig E., Mayer H., 1980. Waldbauliche Terminologie. Wien, Universität für Bodenkultur. Pp. 207.
http://efern.boku.ac.at/forex/wbterm/

564 Wald
(1) Eine Pflanzengesellschaft, die vorwiegend aus Phanerogamen (Bäumen) besteht, die im Reifealter (maturity) mindestens 5 m hoch werden (in subpolaren und subalpinen Zonen auch über 3 m). Neben Bäumen bilden Sträucher, Kräuter und Moose den Pflanzenbestand. Der Wald hat ein besonderes Waldinnenklima. Der natürliche und der bewirtschaftete Wald ist eine Lebensgemeinschaft von Pflanzen und Tieren, deren Zusammenleben durch ökologische Kontrollmechanismen so geregelt wird, daß ein dynamisches, die Erhaltung des Systems sicherndes Gleichgewicht erhalten wird. Dies Gleichgewicht ist kein statistischer Zustand und schließt Katastrophen nicht aus. Geschlossener Wald (dense forest). Der natürliche Überschirmungsgrad ist im Reifestadium 0,6.
Offener Wald (open forest). Der natürliche Überschirmungsgrad ist im Reifestadium 0,3 bis 0,6 (vgl. Gehölz, Gebüsch, Savanne).
(2) Eine bestimmte Landfläche, auf der Forstprodukte und forstliche Infrastrukturleistungen erzeugt werden.
(3) Rechtlich eine Landfläche, die als Wald unter den bestehenden Gesetzen ausgewiesen ist und den entsprechenden gesetzlichen Vorschriften unterliegt.

Source: Brünig E., Mayer H., 1980. Waldbauliche Terminologie. Wien, Universität für Bodenkultur. Pp. 207.
http://efern.boku.ac.at/forex/wbterm/

420 Wald
Vegetationsform, an der Bäume oder Sträucher wesentlich beteiligt sind.

Wird in der *Sanasilva-Inventur gemäss LFI definiert (EAFV 1988).

Source: LWF/SSI (WSL BIRMENSDORF).

582 Walddefinition
Entscheidungsgrundlage zur Abgrenzung von Wald und Nichtwald. Im LFI sind die Kriterien Mindestbreite, minimaler Deckungsgrad und minimale Oberhöhe für den Wald-/ Nichtwaldentscheid massgebend; siehe Anl. LFI S. 32.

Source: Brassel P., Brändli U.B. (Red.), 1999. Schweizerisches Landesforstinventar. Ergebnisse der Zweitaufnahme 1993–1995. Birmensdorf, WSL; Bern, BUWAL; Bern, Haupt. Pp. 442.

23 forêt
Désigne un ensemble de types d'écosystèmes dont le rôle est primordial pour l'ensemble de la biosphère terrestre.

Les forêts se définissent comme des écosystèmes dont la couverture végétale dominante est constituée par des arbres.

En écologie, le terme de forêt concerne des formations végétales dont la frondaison est continue (forêt fermée). Lorsque la couverture est discontinue, on parle de boisements ouverts.

Source: Ramade F., 1993. Dictionnaire encyclopédique de l'écologie et des sciences de l'environnement. Paris, Ediscience international. Pp. 822.

29 forêt
1. Ecologie: Formation végétale ligneuse, ou écosystème, à prédominance d'arbres, comportant en général un couvert relativement dense. Plus particulièrement, formation végétale où prédominent les arbres et autres végétaux ligneux poussant relativement près les uns des autres.
2. Sylviculture/Aménagement forestier: Zone affectée à la production de bois d'oeuvre, et (ou) d'autres produits forestiers, ou que l'on maintient boisée pour en tirer des avantages divers tels que la protection des bassins versants, les bassins de réception, la récréation, etc.

Source: Policy and Economics Directorate Forestry Canada / Direction Générale des politiques et de l'Economie, Forêts Canada, 1992. Silvicultural terms in Canada / Terminologie de la sylviculture au Canada (trad. et adapté de la version anglaise). Ottawa, Science and Sustainable Development Directorate, Forestry Canada. Pp. 74.

34 forêt
Formation végétale ligneuse, ou écosystème, à prédominance d'arbres, comportant en général un couvert relativement dense.

Source: Haddon B.D. (ed), 1988. Forest inventory terms in Canada / Terminologie de l'inventaire des forêts du Canada. 3rd ed.. Ottawa, Canadian Forest Inventory Committee, Forestry Canada. Pp. 113 + 109.

38 forest ecosystem

360 forest ecosystem
An ecological system composed of interacting biotic and abiotic components of the environment in which trees are a major constituent, such that their canopies cover 20% or more of the area (Maini 1991).

Source: Aird P.L., 1994. Conservation for the sustainable development of forests worldwide – a compendium of concepts and terms. Forestry Chronicle 70(6): 666–674.

39 forest health

222 forest health
As a specific condition, refers to a growing forest having many or all of its native species of plants and animals; as a management objective, refers to maintaining or restoring the capacity of a forest to achieve health.

Source: Marcot B.G., Wisdom M.H., Li H.W., Castillo G.C., 1994. Managing for featured, threatened, endangered, and sensitive species and unique habitats for ecosystem sustainability. General Technical Report PNW 329, Portland, OR, USDA Forest Service, Pacific Northwest Research Station, 39 p.

446 forest health
A forest condition that is naturally resilient to damage; characterized by biodiversity, it contains sustained habitat for timber, fish, wildlife, and humans, and meets present and future resource management objectives.

Source: British Columbia Forest Service, March 1997: Glossary of Forestry Terms. Available from the World Wide Web:
http://www.for.gov.bc.ca/pab/publctns/glossary/GLOSSARY.HTM. Cited 01-Oct-98.

91 Gesundheitszustand des Waldes
Fähigkeit des Ökosystems Wald, *Stress in historischem Umfang ohne wesentliche dauernde Änderungen in Zusammensetzung, Struktur und Funktion *Stress zu überstehen und sich an *Stress, der den historischen

Umfang übersteigt, durch Änderungen in Zusammensetzung, Struktur und Funktion anzupassen. Das Ökosystem Wald umfasst alle Sukzessionsstadien.

Source: LWF/SSI (WSL Birmensdorf).

137 Waldzustand
Der Begriff ist nicht selbsterklärend, wurde aber häufig verwendet.
Der Begriff ist hier enger gefasst als im Landesforstinventar (EAFV 1988, S. 223).

40 forest management type

493 Waldform
Grundform des Waldes: Hochwald (gleichförmig, ungleichförmig, plenterartig). Mittelwald, Niederwald und als Spezialformen Selven und Plantagen.

Source: Schütz J.P., Brang P., Bonfils P., Bucher H.U., 1993. Darstellung der Standortansprüche wichtiger Baumarten im Ökogramm und Gesellschaftsanschluss der Baumarten. In: Schmider P., Küper M., Tschander B., Käser B. (eds), Die Waldstandorte im Kanton Zürich. Vdf, Zürich, pp. 254–258.

41 forest management unit

357 FMU
Source: van Bueren E.M.L., Blom E.M., 1997. Hierarchical framework for the formulation of sustainable forest management standards. Leiden, The Netherlands, Backhuys Publishers. Pp. 82.

524 forstliche Betriebsfläche
Die forstliche Betriebsfläche umfaßt alle Flächen des Betriebes, die den Zwecken des forstlichen Betriebes dienen oder keine eigenwirtschaftliche Bedeutung haben (z.B. Hof- und Gebäudeflächen der Dienst- und Werkwohnungen). Sie wird unterteilt in Holzbodenfläche und in Nichtholzbodenfläche. Der Holzboden gliedert sich in Wirtschaftswald und in Wirtschaftswald im außerregelmäßigen Betrieb (a.r.B.).

Source: Brünig E., Mayer H., 1980. Waldbauliche Terminologie. Wien, Universität für Bodenkultur. Pp. 207. http://efern.boku.ac.at/forex/wbterm/

361 forest management unit
An area of forest, planned and managed as an unit to achieve specified management objectives.

Source: Montreal Process Implementation Group, 1998. A framework of regional (sub-national) level criteria and indicators of sustainable forest management in Australia. Canberra, Commonweath of Australia. Pp. 108.

362 forest management unit
An FMU may be defined as a clearly demarcated area of land covered predominantly by forests, managed to a set of explicit objectives and according to a long-term management plan.

Source: van Bueren E.M.L., Blom E.M., 1997. Hierarchical framework for the formulation of sustainable forest management standards. Leiden, The Netherlands, Backhuys Publishers. Pp. 82.

363 forest management unit
A working plan area or any subdivision thereof; mainly, in descending order of importance, working circle, felling series, cutting section.

Source: Ford-Robertson F.C. (ed), 1983. Terminology of forest science, technology, practice and products. 2nd printing with addendum. Washington, D.C., Society of American Foresters. Pp. 370.

526 Bewirtschaftungseinheit
Die Fläche eines Betriebsplanes (working plan) oder eines Teiles davon (Betriebsklasse – working circle, Schlagfolge – felling series). Eine großräumigere, lose (working plan circle) oder straff integrierte (regional management plan, regional forest development plan) Zusammenfassung von Betriebsplaneinheiten zu übergeordneten Planungs- und Vollzugseinheiten wird in zunehmendem Maße zur Verbesserung und Sicherung der nachhaltigen Erfüllung der Rohstofffunktion und Infrastrukturleistung des Waldes durchgeführt.

Source: Brünig E., Mayer H., 1980. Waldbauliche Terminologie. Wien, Universität für Bodenkultur. Pp. 207. http://efern.boku.ac.at/forex/wbterm/

42 forest type

364 forest type
A class in the hierarchy of vegetation classification of forests characterised by the taxonomic and/or structural composition of canopy trees (usually by the dominant species).

Source: Montreal Process Implementation Group, 1998. A framework of regional (sub-national) level criteria and indicators of sustainable forest management in Australia. Canberra, Commonweath of Australia. Pp. 108.

476 forest type
A grouping, or association, of species which comprise plurality of stocking on a given site at the present time. Forest

type is most often selected from the species which make up the forest canopy, not the understory. In cases of recent disturbance, forest type may be determined from the current regeneration. In some regions outside the U.S., "Forest Land Types" are substituted for U.S. Forest Types.

Source: USDA Forest Service, 1998. Forest health monitoring 1998. Field methods guide. Research Triangle Park, NC, USDA Forest Service, National Forest Health Monitoring Program. Pp. [mult. pag.].

535 forest type
A category of forest defined by its vegetation, particularly composition, and/or locality factors, as categorized by each country in a system suitable to its situation.

Source: Criteria and indicators for the conservation and sustainable management of temperate and boreal forests. The Montreal Process, December 1998, 15 p.

43 fragmentation

223 habitat fragmentation
The splitting and isolating of patches of habitat, typically forest cover (but could also apply to grass fields, shrub patches, and other habitats); habitat can be fragmented from natural conditions, such as thin or variable soils, or from forest management activities, such as clearcut logging.

Source: Marcot B.G., Wisdom M.H., Li H.W., Castillo G.C., 1994. Managing for featured, threatened, endangered, and sensitive species and unique habitats for ecosystem sustainability. General Technical Report PNW 329, Portland, OR, USDA Forest Service, Pacific Northwest Research Station, 39 p.

24 fragmentation des habitats
Action par laquelle des phénomènes d'origine naturelle ou anthropique fractionnent les habitats d'un écosystème qui étaient jointifs dans des conditions initiales.
La déforestation due à l'homme, la sécheresse accrue au cours de la dernière glaciation, représentent des exemples de fragmentation des habitats de forêts pluvieuses d'Afrique et d'Amérique tropicale.

Source: Ramade F., 1993. Dictionnaire encyclopédique de l'écologie et des sciences de l'environnement. Paris, Ediscience international. Pp. 822.

281 fragmentation
A term used in conservation biology to refer to the process that converts large areas of relatively uniform vegetation into a mosaic of small patches of vegetation of different age class and wildlife habitat potential across the landscape.

Where old growth forest is fragmented, it may not sustain certain old growth-dependent species even if a significant portion of the landscape still has old growth on it. Such animals or plants may require large patches of mature or old growth forest for their survival.

Source: Kimmins J.P., 1997. Forest ecology. 2nd ed. Upper Saddle River, NJ, Prentice Hall. Pp. 596.

432 fragmentation
The breaking up of extensive landscape features into disjunct, isolated, or semi-isolated patches as a result of land-use changes.

Source: Heywood V.H., Watson R.T., Baste I. (eds), 1995. Global biodiversity assessment. Cambridge, Cambridge University Press. Pp. 1140.

447 fragmentation
The process of transforming large continuous forest patches into one or more smaller patches surrounded by disturbed areas. This occurs naturally through such agents as fire, landslides, windthrow and insect attack. In managed forests timber harvesting and related activities have been the dominant disturbance agents.

Source: British Columbia Forest Service, March 1997: Glossary of Forestry Terms. Available from the World Wide Web:
http://www.for.gov.bc.ca/pab/publctns/glossary/GLOSSARY.HTM. Cited 01-Oct-98.

470 fragmentation
A term used to describe a landscape where areas of forest have been removed in such a way that the remaining forest exists as "islands" of trees in a cutover environment. The major concern with fragmentation is the effect of the loss of contiguous forest cover on species movement and dispersal.

Source: Natural Resources Canada (eds), 1995. The State of Canada's forests 1994. Ottawa, Natural Resources Canada. Pp. 111.

44 functional diversity

365 functional diversity
This can refer to two rather different concepts: the diversity of the ecological functions performed by different species, and the diversity of species performing a given ecological function.

Source: Heywood V.H., Watson R.T., Baste I. (eds), 1995. Global biodiversity assessment. Cambridge, Cambridge University Press. Pp. 1140.

45 genetic diversity

366 genetic diversity
Variation in the genetic composition of individuals within or among species; the heritable genetic variation within and among populations.

Source: Heywood V.H., Watson R.T., Baste I. (eds), 1995. Global biodiversity assessment. Cambridge, Cambridge University Press. Pp. 1140.

46 grain

486 grain size of a landscape
The average, and the variability in, diameter or area of the landscape elements present.

Source: Forman R.T.T., Godron M., 1986. Landscape ecology. New York, Wiley. Pp. 619.

516 grain
The resolution of an image or the minimum area perceived as distinct by an organism.

Source: Farina A., 1998. Principles and methods in landscape ecology. London, Chapman & Hall. Pp. 235.

47 *growth region

565 Wuchsgebiet
Öko-geographisch und vegetationskundlich einheitliches Gebiet, das durch besondere Pflanzengesellschaften, durch Arealgrenzen von Arten und durch besondere edaphische und klimatische Eigenheiten gekennzeichnet ist. Wuchsgebiete können auf Grund geologischer oder topographischer Unterschiede in Wuchslandschaften unterteilt werden; regionale Zusammenfassung mehrerer Einzelwuchsbezirke.

Source: Brünig E., Mayer H., 1980. Waldbauliche Terminologie. Wien, Universität für Bodenkultur. Pp. 207.
http://efern.boku.ac.at/forex/wbterm/

567 Wuchsgebiet
Eine Großlandschaft, die sich durch geomorphologischen Aufbau, Klima und Landschaftsgeschichte von anderen Großlandschaften unterscheidet. Das Wuchsgebiet ist in Wuchsbezirke mit möglichst einheitlichem physiogeographischen Charakter untergliedert.

Source: Wald und Boden. 1996. Schriftenreihe der Sächsischen Landesanstalt für Forsten 7/96.

48 guidelines

367 guidelines
The function of guidelines is to translate criteria and indicators into practical guidance for actions to meet the requirements of criteria and indicators. Guidelines will often be formulated in terms of prescriptions showing how the requirements should be met.

Source: van Bueren E.M.L., Blom E.M., 1997. Hierarchical framework for the formulation of sustainable forest management standards. Leiden, The Netherlands, Backhuys Publishers. Pp. 82.

49 habitat

368 habitat
The space used by an organism, together with the other organisms with which it coexists, and the landscape and climate elements that affect it; the place where an animal or a plant normally lives and reproduces.

Source: Heywood V.H., Watson R.T., Baste I. (eds), 1995. Global biodiversity assessment. Cambridge, Cambridge University Press. Pp. 1140.

538 Habitat
Auf Linné zurückgehender Begriff für den charakteristischen Wohn- oder Standort einer Art. Dieser autökologische Begriff wird oft (besonders in der angelsächsischen Literatur) in synökologischem Sinne als Synonym zu Biotop gebraucht; Umwelt I.e.S.

Source: Brünig E., Mayer H., 1980. Waldbauliche Terminologie. Wien, Universität für Bodenkultur. Pp. 207.
http://efern.boku.ac.at/forex/wbterm/

50 hemeroby

323 hemeroby
(No definition given.)

Source: Sukopp H., 1972. Wandel von Flora und Vegetation in Mitteleuropa unter dem Einfluss des Menschen. Berichte über Landwirtschaft 50: 112–139.

495 Hemerobie
Ein Maß für den menschlichen Kultureinfluß auf Ökosysteme, wobei die Einschätzung des Hemerobiegrades nach dem Ausmaß der Wirkungen derjenigen anthropogenen Einflüsse vorgenommen wird, die der Entwicklung des Systems zu einem Endzustand entgegenstehen.

Ahemerobe Vegetation kann also auch im Zuge der Sukzession einer anthropogenen Pflanzengesellschaft zu einer natürlichen Schlußgesellschaft auf irreversibel veränderten Standorten entstehen.

Source: Kowarik I., 1988. Zum menschlichen Einfluß auf Flora und Vegetation. Theoretische Konzepte und ein Quantifizierungsansatz am Beispiel von Berlin (West). Landschaftsentwicklung und Umweltforschung 56, Berlin, Technische Universität Berlin, 280 S.

496 **Hemerobie**
Bezugsgröße der Hemerobiebewertung ist die ‚potentiell natürliche Waldgesellschaft' (PNWG).

Source: Koch G., Kirchmeir H., 1997. Methodik der Hemerobiebewertung. Österreichische Forstzeitung (1): 24–26.

51 indicator

32 **indicator**
A characteristic of the environment that, when measured, quantifies the magnitude of stress, habitat characteristics, degree of exposure to the stressor, or degree of ecological response to the exposure. Indicators may be of a biotic or abiotic nature.

Source: USDA Forest Service, 1998. Forest health monitoring 1998. Field methods guide. Research Triangle Park, NC, USDA Forest Service, National Forest Health Monitoring Program. Pp. [mult. pag.].

369 **indicator**
In EMAP, characteristics of the environment both abiotic and biotic, that can provide quantitative information on ecological resources. (Revised definition 1993. Preferred term for environmental indicator, deleted 1993.) "In biology, an organism, species, or community whose characteristics show the presence of specific environmental conditions, good or bad" (EPA 1992, 15).

Source: Environmental Monitoring and Assessment Program (EMAP). Master Glossary [online]. Updated 09/15/97 [cited 1998-08-31]. Available from World Wide Web: <http://www.epa.gov/emfjulte/html/glossary.html>

370 **indicator**
An indicator is a quantitative or qualitative parameter which can be assessed in relation to a criterion. It describes in an objectively verifiable and unambiguous way features of the ecosystem or the related social system, or it describes elements of prevailing policy and management conditions and human driven processes indicative of the state of the eco- and social system.

Source: van Bueren E.M.L., Blom E.M., 1997. Hierarchical framework for the formulation of sustainable forest management standards. Leiden, The Netherlands, Backhuys Publishers. Pp. 82.

371 **indicator**
[Indicators] show changes over time for each criterion and demonstrate how well each criterion reaches the objective set for it. A typical indicator in the Helsinki process is a quantitative measure of change, (Helsinki Process).

Source: Granholm H., Vähänen T., Sahlberg S. (compilers), 1996. Background document. Intergovernmental Seminar on Criteria and Indicators for Sustainable Forest Management, August 19–22, 1996, Helsinki. Helsinki, Ministry of Agriculture and Forestry, Intergovernmental Seminar on Criteria and Indicators for Sustainable Forest Management. Pp. 131.

372 **indicator**
[Indicators] are ways of measuring achievements built in criteria, and thus translate the criteria into more direct operational tools, they support the reporting process, and make the reporting internationally credible, (Helsinki Process).

A change in an indicator does not, in itself, tell whether the change has a positive or negative effect. The indicators should therefore be judged on a scale of acceptable "standards of performance", which may vary widely from region to region and from time to time. (Helsinki Process)

Source: Granholm H., Vähänen T., Sahlberg S. (compilers), 1996. Background document. Intergovernmental Seminar on Criteria and Indicators for Sustainable Forest Management, August 19–22, 1996, Helsinki. Helsinki, Ministry of Agriculture and Forestry, Intergovernmental Seminar on Criteria and Indicators for Sustainable Forest Management. Pp. 131.

373 **indicator**
A measure (measurement) of an aspect of the criterion. A quantitative or qualitative variable which can be measured or described and which, when observed periodically, demonstrates trends, (Montreal Process).

Source: Granholm H., Vähänen T., Sahlberg S. (compilers), 1996. Background document. Intergovernmental Seminar on Criteria and Indicators for Sustainable Forest Management, August 19–22, 1996, Helsinki. Helsinki, Ministry of Agriculture and Forestry, Intergovernmental Seminar on Criteria and Indicators for Sustainable Forest Management. Pp. 131.

467 **indicator**
An attribute that characterises another attribute, which is not directly assessed or available.

Source: Köhl M., Päivinen R., 1996. Definition of a system of nomenclature for mapping European forests and for compiling a pan-European forest information system. Luxembourg, Office for Official Publications of the European Communities. Pp. 238.

405 qualitative indicator
A qualitative indicator is expressed as situation, object, or process, and is to be assessed in terms of good/sufficient/unsatisfactory and yes/no.

Source: van Bueren E.M.L., Blom E.M., 1997. Hierarchical framework for the formulation of sustainable forest management standards. Leiden, The Netherlands, Backhuys Publishers. Pp. 82.

406 quantitative indicator
A quantitative indicator is expressed and assessed in terms of amount, numbers, volume, percentages, etc.

Source: van Bueren E.M.L., Blom E.M., 1997. Hierarchical framework for the formulation of sustainable forest management standards. Leiden, The Netherlands, Backhuys Publishers. Pp. 82.

93 Indikator
Eigenschaft der Umwelt, deren Auftreten auf das Vorhandensein, das Ausmass oder die Einwirkungsdauer von Stress hindeutet, oder auf die Reaktion auf diesen Stress. Ein Indikator kann auch auf Habitateigenschaften hindeuten.

Source: LWF/SSI (WSL Birmensdorf).

52 indicator development

374 indicator development
The process through which an indicator is identified, tested, and implemented. A candidate indicator is identified and reviewed by peers before it is selected for further evaluation as a research indicator. Existing data are analyzed, simulation studies are performed with realistic scenarios, and limited field tests are conducted to evaluate the research indicator. In the past, this research indicator was called a "probationary core indicator" or a "development indicator" as it was evaluated in regional demonstration projects. An indicator is considered a core indicator when it is selected for long-term, ecological monitoring as a result of its acceptable performance, demonstrated ability to satisfy the data quality objectives.

Source: Environmental Monitoring and Assessment Program (EMAP). Master Glossary [online]. Updated 09/15/97 [cited 1998-08-31]. Available from World Wide Web: <http://www.epa.gov/emfjulte/html/glossary.html>

53 indicator species

219 ecological indicator species
A species whose population size and trend is assumed to reflect population size and trend of other species associated with the same geographical area and habitats; one type of management indicator species (see that term).

Source: Marcot B.G., Wisdom M.H., Li H.W., Castillo G.C., 1994. Managing for featured, threatened, endangered, and sensitive species and unique habitats for ecosystem sustainability. General Technical Report PNW 329, Portland, OR, USDA Forest Service, Pacific Northwest Research Station, 39 p.

284 indicator species
A species that has a sufficiently consistent association with some environmental condition or other species that its presence can be used to indicate or predict that environmental condition or the potential for that other species.

Source: Kimmins J.P., 1997. Forest ecology. 2nd ed. Upper Saddle River, NJ, Prentice Hall. Pp. 596.

375 indicator species
A plant, animal, or microbial species characteristic of, or that displays characteristic responses to, a specific site, habitat, ecosystem, or environmental condition.

Source: Aird P.L., 1994. Conservation for the sustainable development of forests worldwide – a compendium of concepts and terms. Forestry Chronicle 70(6): 666–674.

376 indicator species
A species whose status provides information on the overall condition of the ecosystem and of other species in that ecosystem. Species which flag changes in biotic or abiotic conditions. They reflect the quality and changes in environmental conditions as well as asects of community composition.

Source: Heywood V.H., Watson R.T., Baste I. (eds), 1995. Global biodiversity assessment. Cambridge, Cambridge University Press. Pp. 1140.

220 management indicator species
A species of fish or wildlife for which a set of management guidelines have been written, chosen for simplifying land management planning; one type of management indicator species is the ecological indicator species.

Source: Marcot B.G., Wisdom M.H., Li H.W., Castillo G.C., 1994. Managing for featured, threatened, endangered, and sensitive species and unique habitats for ecosystem sustainability. General Technical Report PNW 329, Portland, OR, USDA Forest Service, Pacific Northwest Research Station, 39 p.

492 Zeigerpflanzen

Arten, deren Vorkommen oder Fehlen bestimmte Verhältnisse anzeigt wie z.B. Nährstoffarmut oder – reichtum, Bodenfeuchtigkeit, basische oder saure Bodenreaktion, Licht- und Wärmeverhältnisse usw.

Source: Schütz J.P., Brang P., Bonfils P., Bucher H.U., 1993. Darstellung der Standortansprüche wichtiger Baumarten im Ökogramm und Gesellschaftsanschluss der Baumarten. In: Schmider P., Küper M., Tschander B., Käser B. (eds), Die Waldstandorte im Kanton Zürich. Vdf, Zürich, pp. 254–258.

54 insurance value

380 insurance value

The value of biodiversity in maintaining ecosystem functions over a range of environmental conditions.

Source: Heywood V.H., Watson R.T., Baste I. (eds), 1995. Global biodiversity assessment. Cambridge, Cambridge University Press. Pp. 1140.

55 inventory

384 inventorying

The surveying, sorting, cataloguing, quantifying and mapping of entities such as genes, individuals, populations, species, habitats, ecosystems and landscapes or their components, and the synthesis of the resulting information for the analysis of process.

Source: Heywood V.H., Watson R.T., Baste I. (eds), 1995. Global biodiversity assessment. Cambridge, Cambridge University Press. Pp. 1140.

56 isolation

26 isolement

Désigne le fait qu'un groupe d'individus d'une population sédentaire ou une propagule se trouve génétiquement isolé de la population principale, ce qui favorise sa dérive génétique.

Source: Ramade F., 1993. Dictionnaire encyclopédique de l'écologie et des sciences de l'environnement. Paris, Ediscience international. Pp. 822.

57 keystone species

385 keystone species

A species whose loss from an ecosystem would cause a greater than average change in other species populations or ecosystems processes; species that have a disproportionately large effect on other species in a community.

Source: Heywood V.H., Watson R.T., Baste I. (eds), 1995. Global biodiversity assessment. Cambridge, Cambridge University Press. Pp. 1140.

517 keystone species

Species that shapes the habitat in which it lives and allows the presence of other species.

Source: Farina A., 1998. Principles and methods in landscape ecology. London, Chapman and Hall. Pp. 235.

28 espèce clef

A l'intérieur d'une communauté existent des espèces qui, sans être nécessairement les plus abondantes – ou les plus spectaculaires par leur taille – jouent un rôle essentiel car elles assurent la structuration de la communauté et même conditionnent la richesse spécifique de cette dernière.

Source: Ramade F., 1993. Dictionnaire encyclopédique de l'écologie et des sciences de l'environnement. Paris, Ediscience international. Pp. 822.

58 landscape

428 landscape

An extensive area of terrain.

Source: Freedman B., 1989. Environmental ecology. The impacts of pollution and other stresses on ecosystem structure and function. San Diego, Academic Press. Pp. 424.

448 landscape

The fundamental traits of a specific geographic area, including its biological composition, physical environment and anthropogenic or social patterns.

Source: British Columbia Forest Service, March 1997: Glossary of Forestry Terms. Available from the World Wide Web: http://www.for.gov.bc.ca/pab/publctns/glossary/GLOSSARY.HTM. Cited 01-Oct-98.

471 landscape

Areas of land that are distinguished by differences in landforms, vegetation, land use, and aesthetic characteristics.

Source: Natural Resources Canada (eds), 1995. The State of Canada's forests 1994. Ottawa, Natural Resources Canada. Pp. 111.

484 landscape
A heterogeneous land area composed of a cluster of interacting ecosystems that are repeated in similar form throughout. Landscapes vary in size, down to a few kilometers in diameter.

Source: Forman R.T.T., Godron M., 1986. Landscape ecology. New York, Wiley. Pp. 619.

514 landscape
There are several definitions of landscape:
 1. "the total character of a region" (von Humboldt);
 2. "landscapes dealt with in their totality as physical, ecological and geographical entities, integrating all natural and human ("caused") patterns and processes..." (Naveh 1987);
 3. "landscape as a heterogeneous land area composed of a cluster of interacting ecosystems that is repeated in similar form throughout" (Forman and Godron 1986);
 4. "a particular configuration of topography, vegetation cover, land use and settlement pattern which delimits some coherence of natural and cultural processes and activities" (Green et al. 1996);
 5. Haber has defined the landscape as "a piece of land which we perceive comprehensively around us, without looking closely at single components, and which looks familiar to us" (pers. com. 1996).

Source: Farina A., 1998. Principles and methods in landscape ecology. London, Chapman and Hall. Pp. 235.

542 Landschaft
Ein Teil der Erdoberfläche, der durch die besondere Konstellation der Landschaftselemente (Topographie, Klima, Boden, bio-geographische Verhältnisse) geprägt ist und sich hierdurch gegenüber anderen Landschaften unterscheidet. Das Zusammenwirken der Landschaftselemente bewirkt strukturell die Landschaftsstruktur, visuell das Landschaftsbild und funktionell den Landschaftshaushalt. Man unterscheidet die ursprüngliche Naturlandschaft und die vom Menschen gestaltete Kulturlandschaft.
 Im allgemeinen Sinne spricht man von Agrarlandschaft, Seenlandschaft, Industrielandschaft u.a. Hinter dieser äußeren Gestalt steht ein inneres Gefüge, wobei Naturausstattung, die Nutzungsformen, die Tätigkeit des Menschen, die Bauwerke u.a. in einem engen räumlichen Zusammenhang stehen und aufeinander in vielfältiger Weise einwirken.

Source: Brünig E., Mayer H., 1980. Waldbauliche Terminologie. Wien, Universität für Bodenkultur. Pp. 207.
http://efern.boku.ac.at/forex/wbterm/

55 paysage
Portion structurée de territoire observable globalement à partir d'un point donné, comprenant un ensemble d'éléments naturels géomorphologiques, hydrologiques (éventuellement), végétaux (formations végétales), et/ou d'origine artificielle liés à l'action humaine: terres cultivées, constructions, voies de communiation, etc.

Source: Delpech R., Dume G., Galmiche P., Timbal J., 1985. Vocabulaire: typologie des stations forestières. Paris, Institut pour le Développement Forestier. Pp. 243.

59 landscape structure

485 landscape structure
The distribution of energy, materials, and species in relation to the sizes, shapes, numbers, kinds, and configurations of landscape elements or ecosystems.

Source: Forman R.T.T., Godron M., 1986. Landscape ecology. New York, Wiley. Pp. 619.

60 *man-made

482 naturferner Bestand
Bestand mit mittlerem, im allgemeinen tragbarem Anteil an standortfremden Baumarten und erkennbaren natürlichen Merkmalen.

Source: Schütz J.P., Brang P., Bonfils P., Bucher H.U., 1993. Darstellung der Standortansprüche wichtiger Baumarten im Ökogramm und Gesellschaftsanschluss der Baumarten. In: Schmider P., Küper M., Tschander B., Käser B. (eds), Die Waldstandorte im Kanton Zürich. Vdf, Zürich, pp. 254–258.

497 naturfern
Vom Menschen geschaffen und vollständig von ihm abhängig; Synonym zu anthropogen. Beispiel: landwirtschaftliche Kulturen.

Source: Schaefer M., 1992. Wörterbuch der Biologie. Ökologie. 3. Auflage. Jena, Gustav Fischer Verlag. Pp. 433.

61 matrix

209 matrix
The most extensive and most connected landscape element type present, which plays the dominant role in landscape functioning. Also, a landscape element surrounding a patch.

Source: Forman R.T.T., Godron M., 1986. Landscape ecology. New York, Wiley. Pp. 619.

62 minimum viable population

386 minimum viable population
The smallest isolated population having a good chance of surviving for a given number of years despite the foreseeable effects of demographic, environmental, and genetic events and natural catastrophes. (The probability of persistence and the time of persistence are often taken to be 99% and 1000 years, respectively.)

Source: Heywood V.H., Watson R.T., Baste I. (eds), 1995. Global biodiversity assessment. Cambridge, Cambridge University Press. Pp. 1140.

63 monitoring

84 monitoring
The collection of information over time to determine the effects of resource management and to identify changes in natural systems.

Source: USDA Forest Service, 1998. Forest health monitoring 1998. Field methods guide. Research Triangle Park, NC, USDA Forest Service, National Forest Health Monitoring Program. Pp. [mult. pag.].

387 monitoring
The intermittent (regular or irregular) surveillance to ascertain the extent of compliance with a predetermined standard or degree of deviation from an expected norm.

Source: Hellawell J.M., 1991. Development of a rationale for monitoring. In: B. Goldsmith (ed), Monitoring for conservation and ecology. London, Chapman and Hall, pp. 1–14.

64 multiple use

188 multiple-use forestry
Any practice of forestry fulfilling two or more objects of management, more particularly in forest utilization, e.g. production of both wood and pasture.

Source: Ford-Robertson F.C. (ed), 1983. Terminology of forest science, technology, practice and products. 2nd printing with addendum. Washington, D.C., Society of American Foresters. Pp. 370.

543 Mehrzwecknutzung
Die Erzeugung und Nutzung von mehr als einem Forstprodukt und bzw. oder mehr als einer Infrastrukturleistung, entweder gleichzeitig auf einer Forstfläche oder auf getrennten Flächen innerhalb einer forstlichen Betriebseinheit. Gebirgswälder meist mit kombinierter Ertrags- und Schutzfunktion.

Source: Brünig E., Mayer H., 1980. Waldbauliche Terminologie. Wien, Universität für Bodenkultur. Pp. 207. http://efern.boku.ac.at/forex/wbterm/

189 multiple-purpose forestry
Source: Ford-Robertson F.C. (ed), 1983. Terminology of forest science, technology, practice and products. 2nd printing with addendum. Washington, D.C., Society of American Foresters. Pp. 370.

388 multiple use
Two or more activities, such as hiking, hunting, or logging, occurring together on the same area either (a) intermixed, (b) confined to separate zones, or (c) in sequence.

Source: Aird P.L., 1994. Conservation for the sustainable development of forests worldwide – a compendium of concepts and terms. Forestry Chronicle 70(6): 666–674.

65 native forest

389 native forest
Any locally indigenous forest community containing species and habitats normally associated with that community.

Source: Montreal Process Implementation Group, 1998. A framework of regional (sub-national) level criteria and indicators of sustainable forest management in Australia. Canberra, Commonweath of Australia. Pp. 108.

66 natural forest

391 natural forest
A forest that has evolved and reproduced itself naturally from organisms previously established, and that has not been significantly altered by human activity. A natural forest may include, but is not equivalent to, an "old-growth forest".

Source: Aird P.L., 1994. Conservation for the sustainable development of forests worldwide – a compendium of concepts and terms. Forestry Chronicle 70(6): 666–674.

392 natural forest
A forest which has evolved as a sequence of natural succession but is still showing anthropogenic influences. Also, forests that have developed from unmanaged pastures or from fallow land. Often natural parks are included in this category.

Source: Loiskekosti M., Halko L. (eds), 1993. Ministerial Conference on the Protection of Forests in Europe, 16–17 June 1993 in Helsinki. European list of criteria and most suitable quantitative indicators. Helsinki, Ministry of Agriculture and Forestry. Pp. 20.

472 open forests
Proposed name for the natural forests commonly found in northern Canada. These forests are a mixture of wetlands and small trees, occasionally interspersed with highly productive forests.

Source: Natural Resources Canada (eds), 1995. The State of Canada's forests 1994. Ottawa, Natural Resources Canada. Pp. 111.

545 Naturwald
Ohne Eingriffe des Menschen entstandener Wald. Er entspricht in seinen Strukturmerkmalen etwa dem Klimaxwald, wenn die natürliche Entwicklung durch "Naturkatastrophen" (Wind, Schnee, Feuer, Überschwemmung) nicht gehemmt wurde.

Source: Brünig E., Mayer H., 1980. Waldbauliche Terminologie. Wien, Universität für Bodenkultur. Pp. 207.
http://efern.boku.ac.at/forex/wbterm/

578 Naturwald
Wald, der nur soweit beeinflusst ist, dass sich Baumartenmischung und Struktur innerhalb einer Baumgeneration in den ursprünglichen Zustand zurückentwickeln können (Wasser und Frehner 1996).

Source: Brassel P., Brändli U.B. (Red.), 1999. Schweizerisches Landesforstinventar. Ergebnisse der Zweitaufnahme 1993–1995. Birmensdorf, WSL; Bern BUWAL; Bern, Haupt. Pp. 442.

67 natural landscape

211 natural landscape
An area where human effects, if present, are not ecologically significant to the landscape as a whole.

Source: Forman R.T.T., Godron M., 1986. Landscape ecology. New York, Wiley. Pp. 619.

68 natural regeneration

178 natural regeneration
Renewal of a tree crop by natural seeding, sprouting, suckering, or layering.

Source: Policy and Economics Directorate Forestry Canada / Direction Générale des politiques et de l'Economie, Forêts Canada, 1992. Silvicultural terms in Canada / Terminologie de la sylviculture au Canada (trad. et adapté de la version anglaise). Ottawa, Science and Sustainable Development Directorate, Forestry Canada. Pp. 74.

186 natural regeneration
Renewal [of a tree crop] by self-sown seed or by vegetative means (regrowth), e.g. coppicing, root suckers, lignotubers, as also the resultant crop.

Forests include special kinds such as industrial forests, non-industrial private forests, plantations, public forests, protection forests, urban forests, as well as parks and wilderness.

Source: Ford-Robertson F.C. (ed), 1983. Terminology of forest science, technology, practice and products. 2nd printing with addendum. Washington, D.C., Society of American Foresters. Pp. 370.

530 natural regeneration
Re-establishment of a forest stand by natural means, i.e. by natural seeding or vegetation regeneration. It may be assisted by human intervention, e.g. by scarification or fencing to protect against wildlife damage or domestic animal grazing.

Source: UN-ECE/FAO, 1997. UN-ECE/FAO Temperate and Boreal Forest Resources Assessment 2000. Terms and definitions. New York and Geneva, United Nations. Pp. 13.

513 Aufschlag
Bestand im jugendlichen Alter, der aus nicht flugfähigen Samen durch Naturverjüngung entstanden ist (z.B. Eiche, Buche).

Source: Brünig E., Mayer H., 1980. Waldbauliche Terminologie. Wien, Universität für Bodenkultur. Pp. 207.
http://efern.boku.ac.at/forex/wbterm/

471 Naturverjüngung
Natürlich durch Ansamung oder durch vegetative Vermehrung entstandene Verjüngung, im Gegensatz zur Kunstverjüngung.

Source: Brassel P., Brändli U.B. (Red.), 1999. Schweizerisches Landesforstinventar. Ergebnisse der Zweitaufnahme 1993–1995. Birmensdorf, WSL; Bern, BUWAL; Bern, Haupt. Pp. 442.

486 Naturverjüngung
Natürlich aufkommende, d.h. weder gesäte noch gepflanzte Verjüngung.

Source: Schütz J.P., Brang P., Bonfils P., Bucher H.U., 1993. Darstellung der Standortansprüche wichtiger

Baumarten im Ökogramm und Gesellschaftsanschluss der Baumarten. In: Schmider P., Küper M., Tschander B., Käser B. (eds), Die Waldstandorte im Kanton Zürich. Vdf, Zürich, pp. 254–258.

544 Naturverjüngung
(1) Begründung eines Bestandes durch Selbstansamung oder vegetative Vermehrung von einem Altbestand aus.
(2) Durch Selbsansamung oder vegetative Vermehrung entstandener junger Bestand.

Source: Brünig E., Mayer H., 1980. Waldbauliche Terminologie. Wien, Universität für Bodenkultur. Pp. 207.
http://efern.boku.ac.at/forex/wbterm/

33 régénération naturelle
Renouvellement naturel d'un peuplement forestier par voie de semences, par des rejets, par drageonnement ou par marcottage.

Source: Policy and Economics Directorate Forestry Canada / Direction Générale des politiques et de l'Economie, Forêts Canada, 1992. Silvicultural terms in Canada / Terminologie de la sylviculture au Canada (trad. et adapté de la version anglaise). Ottawa, Science and Sustainable Development Directorate, Forestry Canada. Pp. 74.

69 near-natural

485 naturnaher Bestand
Bestand mit kleinem Anteil an standortfremden Baumarten. Er besteht zum grössten Teil aus standortheimischen Baumarten mit einem weitgehend naturnahem Beziehungsgefüge.

Source: Schütz J.P., Brang P., Bonfils P., Bucher H.U., 1993. Darstellung der Standortansprüche wichtiger Baumarten im Ökogramm und Gesellschaftsanschluss der Baumarten. In: Schmider P., Küper M., Tschander B., Käser B. (eds), Die Waldstandorte im Kanton Zürich. Vdf, Zürich, pp. 254–258.

498 naturnah
Ohne direkten Einfluß des Menschen entstanden und in geringem Maße von ihm beeinflußt.

Source: Schaefer M., 1992. Wörterbuch der Biologie. Ökologie. 3. Auflage. Jena, Gustav Fischer Verlag. Pp. 433.

70 niche

282 fundamental niche
The geographical range and habitat a species can occupy and the ecological (functional) role it can fulfill in the ecosystem (i.e., its ecological niche) as determined by its genetically determined tolerances and requirements. The niche a species can occupy in the absence of competition or other antagonistic interactions with other species.

Source: Kimmins J.P., 1997. Forest ecology. 2nd ed. Upper Saddle River, NJ, Prentice Hall. Pp. 596.

181 microsite
The ultimate unit of the habitat, i.e., the specific spot occupied by an individual organism. By extension, the more or less specialized relationships existing between an organism and its environment.

Source: Policy and Economics Directorate Forestry Canada / Direction Générale des politiques et de l'Economie, Forêts Canada, 1992. Silvicultural terms in Canada / Terminologie de la sylviculture au Canada (trad. et adapté de la version anglaise). Ottawa, Science and Sustainable Development Directorate, Forestry Canada. Pp. 74.

193 niche
Role, function, or place of organism.

Source: Colinvaux P.A., 1993. Ecology 2. 2nd ed. New York, Wiley. Pp. 688.

203 niche
A confused word used in several different ways over the last half-century. Sometimes it refers to the habitat of an organism or species, and sometimes it refers to the role the organism or species plays in the larger community of which it is part. More recent definitions refer to the resource base upon which the organism or species is characteristically dependent.

Source: Allen T.F.H., Starr T.B., 1982. Hierarchy. Perspectives for ecological complexity. Chicago, University of Chicago Press. Pp. 310.

291 niche
The geographical range and habitat a species can or does occupy, and the ecological (functional) role it can or does fulfill in an ecosystem. The functional, adaptational, and distributional characteristics of a species. A species has a genetically controlled fundamental niche, but it generally occupies only a subset of this (the realized niche) because of competition and other antagonistic community interactions.

Source: Kimmins J.P., 1997. Forest ecology. 2nd ed. Upper Saddle River, NJ, Prentice Hall. Pp. 596.

393 niche
The unique environment used to sustain the existence of an organism or species.

Source: Aird P.L., 1994. Conservation for the sustainable development of forests worldwide – a compendium of concepts and terms. Forestry Chronicle 70(6): 666–674.

394 niche
The place occupied by a species in its ecosystem and its role: where it lives, what it feeds on and when it performs all its activities.

Source: Heywood V.H., Watson R.T., Baste I. (eds), 1995. Global biodiversity assessment. Cambridge, Cambridge University Press. Pp. 1140.

194 niche space
Environmental parameters defining a niche, or the resource flux required for an individual to survive and reproduce.

Source: Colinvaux P.A., 1993. Ecology 2. 2nd ed. New York, Wiley. Pp. 688.

297 realized niche
The geographical area and habitat a species can occupy, and the functional role it can play in an ecosystem, in face of competition and other antagonistic interactions from other species.

Source: Kimmins J.P., 1997. Forest ecology. 2nd ed. Upper Saddle River, NJ, Prentice Hall. Pp. 596.

547 Nische
Ökologische Nische, vielseitige Definition.
(1) Bezeichnung für das Wirkungsfeld, die Rolle, die Stellung einer Art in einem Ökosystem. Beziehungssystem zwischen Tier und Umwelt.
(2) Von einigen Autoren auch als Synonym zu Minimalumwelt einer Art gebraucht; dann ist auch die räumliche Komponente mit eingeschlossen; Habitat, Monotop, minimale Umwelt eines Tieres, ökologisch besonders geeigneter Raum, ökologischer Aufenthaltsraum, Nahrungsnische.
(3) Die Spezialisierung einer Art innerhalb einer Gesellschaft; die besondere Weise, in der eine Art sich in Bezug auf andere Arten der gleichen Lebensform und auf die sonstigen biotischen und abiotischen Umweltbedingungen innerhalb der Biozönose einstellt.

Source: Brünig E., Mayer H., 1980. Waldbauliche Terminologie. Wien, Universität für Bodenkultur. Pp. 207. http://efern.boku.ac.at/forex/wbterm/

71 norm

396 norm
A norm is the reference value of the indicator and is established for use as a rule or a basis for comparison. By comparing the norm with the actual measured value, the result demonstrates the degree of fulfilment of a criterion and of compliance with a principle.

Source: van Bueren E.M.L., Blom E.M., 1997. Hierarchical framework for the formulation of sustainable forest management standards. Leiden, The Netherlands, Backhuys Publishers. Pp. 82.

72 ownership

439 ownership
A classification of forest land based on the legal owner at the time of the current inventory. Also indicates private lands leased to forest industry. Individual ownerships are logically organized into ownership groups and classes for reporting purposes [national forest, other public, forest industry, other private].

Source: Southern Region Forest Inventory and Analysis, Forest Inventory – Definitions of Terms (no date). Available from the World Wide Web: http:// www.srsfia.usfs.msstate.edu/fidef2.htm. Cited 01-Oct-98.

531 private ownership (in)
Forest/other wooded land owned by individuals, families, co-operatives and corporations which may be engaged in agriculture or other occupations as well as forestry; private forest enterprises and industries; private corporations and other institutions (religious and educational institutions, pension and investment funds, nature conservation societies, etc.).

Source: UN-ECE/FAO, 1997. UN-ECE/FAO Temperate and Boreal Forest Resources Assessment 2000. Terms and definitions. New York and Geneva, United Nations. Pp. 13.

532 public ownership (in)
Forest/other wooded land belonging to the State or other public bodies.

Source: UN-ECE/FAO, 1997. UN-ECE/FAO Temperate and Boreal Forest Resources Assessment 2000. Terms and definitions. New York and Geneva, United Nations. Pp. 13.

73 parameter

379 input parameter

An input parameter is an object, capacity, or intention, put in, or taken in, or operated on by any human driven process.

Source: van Bueren E.M.L., Blom E.M., 1997. Hierarchical framework for the formulation of sustainable forest management standards. Leiden, The Netherlands, Backhuys Publishers. Pp. 82.

398 output parameter

An output parameter is the actual or desired result of a management process which describes the state or capacity of the ecosystem, the state of a physical component or the state of the related social system or its components.

An output parameter may also be referred to as an output or performance parameter.

Source: van Bueren E.M.L., Blom E.M., 1997. Hierarchical framework for the formulation of sustainable forest management standards. Leiden, The Netherlands, Backhuys Publishers. Pp. 82.

466 parameter

A value, known or unknown, applied to a population rather than a sample, which is used to define a statistical model, usually a theoretical distribution.

Source: Köhl M., Päivinen R., 1996. Definition of a system of nomenclature for mapping European forests and for compiling a pan-European forest information system. Luxembourg, Office for Official Publications of the European Communities. Pp. 238.

479 parameter

Any of the numerical constants which appear in a mathematical expression of relationship among variables. A parameter is called dimensionless if the related variables are expressed in the same physical units, e.g. $g/m^2/yr$.

Source: Allen T.F.H., Starr T.B., 1982. Hierarchy. Perspectives for ecological complexity. Chicago, University of Chicago Press. Pp. 310.

404 process parameter

A process parameter is the management process or a component of the management process, or other human action, describing human activities and not the result of the activity (planning process, field operations).

Source: van Bueren E.M.L., Blom E.M., 1997. Hierarchical framework for the formulation of sustainable forest management standards. Leiden, The Netherlands, Backhuys Publishers. Pp. 82.

74 patch

212 patch

A nonlinear surface area differing in appearance from its surrounding.

Source: Forman R.T.T., Godron M., 1986. Landscape ecology. New York, Wiley. Pp. 619.

75 patch dynamics

506 patch dynamics

The concept of communities as consisting of a mosaic of patches within which abiotic disturbances and biotic interactions proceed.

Source: Begon M., Harper J.L., Townsend C.R., 1996. Ecology: Individuals, populations and communities. Third edition. Oxford, Blackwell. Pp. 1068.

76 patch turnover

213 patch turnover

The rate of appearance and disappearance of patches.

Source: Forman R.T.T., Godron M., 1986. Landscape ecology. New York, Wiley. Pp. 619.

77 plantation

162 plantation

Forest stands established by planting or/and seeding in the process of afforestation or reforestation. They are either:
  – of introduced species (all planted stands), or
  – intensively managed stands of indigenous species which meet all the following criteria: one or two species at plantation, even age class, regular spacing. Excludes: Stands which were established as plantation but which have been without intensive management for a significant period of time. These should be considered seminatural.

Source: UN-ECE FAO Temperate and Boreal Forest Resources
Assessment 2000. Terms and Definitions. July 1997. GE. 97-2223I. 13 p.

32 plantation

Le résultat d'une plantation:
  1. Au sens large, action de planter des arbres par ensemencement direct ou par mise en terre de plants ou de boutures.

2. Au sens strict, action de créer une forêt en plantant de jeunes plants et non pas par ensemencement direct.

Source: Policy and Economics Directorate Forestry Canada / Direction Générale des politiques et de l'Economie, Forêts Canada, 1992. Silvicultural terms in Canada / Terminologie de la sylviculture au Canada (trad. et adapté de la version anglaise). Ottawa, Science and Sustainable Development Directorate, Forestry Canada. Pp. 74.

183 plantation
A forest crop established artificially, either by sowing or planting.

Source: Policy and Economics Directorate Forestry Canada / Direction Générale des politiques et de l'Economie, Forêts Canada, 1992. Silvicultural terms in Canada / Terminologie de la sylviculture au Canada (trad. et adapté de la version anglaise). Ottawa, Science and Sustainable Development Directorate, Forestry Canada. Pp. 74.

190 plantation
A forest crop or stand raised artificially, either by sowing or planting.

Source: Ford-Robertson F.C. (ed), 1983. Terminology of forest science, technology, practice and products. 2nd printing with addendum. Washington, D.C., Society of American Foresters. Pp. 370.

399 plantation
Intensively managed stand of trees of either native or exotic species, created by the regular placement of seedlings or seed.

Source: Montreal Process Implementation Group, 1998. A framework of regional (sub-national) level criteria and indicators of sustainable forest management in Australia. Canberra, Commonweath of Australia. Pp. 108.

78 population

473 population
A group of organisms of the same species inhabiting a particular geographical area at a particular time.

Source: Natural Resources Canada (eds), 1995. The State of Canada's forests 1994. Ottawa, Natural Resources Canada. Pp. 111.

548 Population
(l) In der Statistik eine Menge, deren Elemente mindestens in einem MerkmalGemeinsamkeiten aufweisen.

(2) In der Genetik eine Gemeinschaft von gemischterbigen Pflanzen oder Tieren, die auf einem begrenzten Raum leben, sich durch Fremdbefruchtung vermehren und an einem gemeinsamen "gen-pool" teilhaben.
(3) In der Ökologie eine Gemeinschaft von Individuen der gleichen Art, die in gegenseitiger Wechselbeziehung stehen und die sich in einem einheitlichen Raum (z.B. Biotop) befinden und im allgemeinen durch mehrere Generationen genetische Kontinuität zeigen.

Source: Brünig E., Mayer H., 1980. Waldbauliche Terminologie. Wien, Universität für Bodenkultur. Pp. 207. http://efern.boku.ac.at/forex/wbterm/

79 population vulnerability analysis

400 population viability analysis
Source: Gilpin M.E., Soulé M.E., 1986. Minimum viable populations: processes of species extinction. In: M.E. Soulé (ed), Conservation biology: the science of scarcity and diversity. Sunderland, Mass., Sinauer Associates, Inc.

401 population viability analysis
A comprehensive analysis of the many environmental and demographic factors that affect survival of a population, usually small.

Source: Heywood V.H., Watson R.T., Baste I. (eds), 1995. Global biodiversity assessment. Cambridge, Cambridge University Press. Pp. 1140.

402 population vulnerability analysis
An analytical technique that estimates the minimum viable population of a species required to sustain its existence.

Source: Gilpin M.E., Soulé M.E., 1986. Minimum viable populations: processes of species extinction. In: M.E. Soulé (ed), Conservation biology: the science of scarcity and diversity. Sunderland, Mass., Sinauer Associates, Inc.

509 population vulnerability analysis
An analysis, generally applied to populations or species in danger of extinction, of the population's chances of extinction.

Source: Begon M., Harper J.L., Townsend C.R., 1996. Ecology: Individuals, populations and communities. Third edition. Oxford, Blackwell. Pp. 1068.

510 PVA
Source: Begon M., Harper J.L., Townsend C.R., 1996. Ecology: Individuals, populations and communities. Third edition. Oxford, Blackwell. Pp. 1068.

80 primary forest

559 Urwald

Naturwald mit natürlichem Bestandesaufbau ohne jeden anthropogenen Einfluß in Vergangenheit und Gegenwart; primärer Urwald. Sekundärer Urwald: natürlicher (naturnaher) Waldzustand, der heute keinen offensichtlichen anthropogenen Einfluß mehr erkennen läßt, bzw. frühere menschliche Einwirkungen nicht oder nur in unwesentlichen Merkmalen aufweist.

Source: Brünig E., Mayer H., 1980. Waldbauliche Terminologie. Wien, Universität für Bodenkultur. Pp. 207.
http://efern.boku.ac.at/forex/wbterm/

81 principle

403 principle

A principle is a fundamental law or rule, serving as a basis for reasoning and action. Principles have the character of an objective or attitude concerning the function of the forest ecosystem or concerning a relevant aspect of the social system that interacts with the ecosystem. Principles are explicit elements of a goal e.g. sustainable forest management or well managed forests.

Source: van Bueren E.M.L., Blom E.M., 1997. Hierarchical framework for the formulation of sustainable forest management standards. Leiden, The Netherlands, Backhuys Publishers. Pp. 82.

82 provenance

296 provenance

The geographical location and its physical features where a particular genotype evolved. Generally used to describe the location of plant seed collections, and described by latitude, longitude, elevation, aspect, and climate.

Source: Kimmins J.P., 1997. Forest ecology. 2nd ed. Upper Saddle River, NJ, Prentice Hall. Pp. 596.

474 provenance

The geographical area or place of origin of a collection of genetic material (generally in the form of seed, pollen or cuttings) for which the process of natural selection has resulted in some common or shared population characteristics.

Source: Natural Resources Canada (eds), 1995. The State of Canada's forests 1994. Ottawa, Natural Resources Canada. Pp. 111.

83 rare species

407 rare species

Taxa with small world populations that are not at present "endangered" or "vulnerable" but at risk. These taxa are usually localised within restricted geographical areas or habitats or are thinly scattered over a more extensive range.

Source: UCN Conservation Monitoring Centre, Cambridge U.K. (compiler), 1994. IUCN red list of threatened animals. Gland, IUCN. Cited in Loiskekosti M., Halko L. (eds), 1993. Ministerial Conference on the Protection of Forests in Europe, 16–17 June 1993 in Helsinki. European list of criteria and most suitable quantitative indicators. Helsinki, Ministry of Agriculture and Forestry. Pp. 20.

84 rarity

526 rarity

See Prevalence (of abundance) and Intensity of abundance.

Source: Begon M., Harper J.L., Townsend C.R., 1996. Ecology: Individuals, populations and communities. Third edition. Oxford, Blackwell. Pp. 1068.

25 rareté

Propriété de nombreuses espèces végétales ou animales d'être représentées par des populations très peu nombreuses. On peut distinguer deux types d'espèces rares. Les premières peuvent se rencontrer en un assez grand nombre d'habitats géographiquement éloignés, mais elles présentent toujours une densité de population très faible. A l'opposé, il existe un second type d'espèces très sténœciques et inféodées à des niches écologiques, elles-mêmes peu fréquentes. Dans ce cas, ces espèces peuvent avoir dans leur habitat une forte densité mais ne se rencontrent qu'en un nombre très faible de biotopes.

Source: Ramade F., 1993. Dictionnaire encyclopédique de l'écologie et des sciences de l'environnement. Paris, Ediscience international. Pp. 822.

85 regenerated patch

214 regenerated patch

An area that becomes free of disturbance within a chronically disturbed matrix.

Source: Forman R.T.T., Godron M., 1986. Landscape ecology. New York, Wiley. Pp. 619.

215 remnant patch

An area remaining from a former large landscape element and now surrounded by a disturbed area.

Source: Forman R.T.T., Godron M., 1986. Landscape ecology. New York, Wiley. Pp. 619.

86 regeneration

533 regeneration
Re-establishment of a forest land by natural or artificial means following the removal of the previous stand by felling or as a result of natural causes, e.g. fire or storm.

Source: UN-ECE/FAO, 1997. UN-ECE/FAO Temperate and Boreal Forest Resources Assessment 2000. Terms and definitions. New York and Geneva, United Nations. Pp. 13.

501 Verjüngung
Bestandesbegründung (Vorgang): Schlagen der alten Bäume zur Einleitung der Jungwaldphase.
Jungwald (Zustand): Ansamung, Keimlinge und junge Bäumchen in der Krautschicht.

Source: Schütz J.P., Brang P., Bonfils P., Bucher H.U., 1993. Darstellung der Standortansprüche wichtiger Baumarten im Ökogramm und Gesellschaftsanschluss der Baumarten. In: Schmider P., Küper M., Tschander B., Käser B. (eds), Die Waldstandorte im Kanton Zürich. Vdf, Zürich, pp. 254–258.

562 Verjüngung
(1) Die Population der Verjüngung (recruitment). Sie kann aus Sämlingen (seedlings) bis zum Ende des Keimlingsstadium aus Lohden (saplings) von ca 0.5–2.5 m Höhe und bis etwa 5–10 cm Durchmesser sowie aus Stangen größerer Dimensionen (larges poles) bestehen.
(2) Die Summe der natürlichen Ereignisse und waldbaulichen Maßnahmen zur Erzielung und Förderung der Verjüngung (regeneration).
(3) Auf künstlichem oder natürlichem Wege wiederbegründeter Bestand im jugendlichen Alter, natürliche oder künstliche Bestandesbegründung.

Source: Brünig E., Mayer H., 1980. Waldbauliche Terminologie. Wien, Universität für Bodenkultur. Pp. 207. http://efern.boku.ac.at/forex/wbterm/

87 resilience

278 elasticity
The speed with which an ecosystem returns to its original condition following disturbance. Also called resilience. A measure of ecosystem stability.

Source: Kimmins J.P., 1997. Forest ecology. 2nd ed. Upper Saddle River, NJ, Prentice Hall. Pp. 596.

468 ecological or ecosystem resilience
Ecological resilience can be defined in two ways. The first is a measure of the magnitude of disturbance that can be absorbed before the (eco)system changes its structure by changing the variables and processes that control behaviour. The second, a more traditional meaning, is as a measure of resistance to disturbance and the speed of return to the equilibrium state of an ecosystem.

Source: Heywood V.H., Watson R.T., Baste I. (eds), 1995. Global biodiversity assessment. Cambridge, Cambridge University Press. Pp. 1140.

88 riparian

449 riparian
An area of land adjacent to a stream, river, lake or wetland that contains vegetation that, due to the presence of water, is distinctly different from the vegetation of adjacent upland areas.

Source: British Columbia Forest Service, March 1997: Glossary of Forestry Terms. Available from the World Wide Web: http://www.for.gov.bc.ca/pab/publctns/glossary/GLOSSARY.HTM. Cited 01-Oct-98.

523 riparian area
There is no standard definition of a riparian area. (...)The following two definitions desribe a natural riparian ecosystem.
The Willamette National Forest's definition of riparian area is: "The aquatic ecosystem and the portions of the adjacent terrestrial ecosystem that directly affect or are affected by the aquatic environment. This includes streams, rivers, and lakes and their adjacent side channels, floodplains and wetlands. The riparian area includes portions of hillslope that serve as streamside habitats for wildlife" (Gregory and Ashkenas 1990).
The definition in the British Columbia Forest Practices Code is: "The land adjacent to the normal high water line in a stream, river, or lake, extending to the portion of land that is influenced by the presence of the adjacent ponded or channeled water. Riparian areas typically exemplify a rich and diverse vegetative mosaic reflecting the influence of available surface water". (B.C. Ministry of Forests 1994).
B.C. Ministry of Forests. 1994. British Columbia Forest Practices Code: Standards. 216 pp.
Gregrory S., Ashkenas L. 1990. Riparian management guide. Willamette National Forest. 120 pp.

Source: Voller J., 1998. Riparian areas and wetlands. In: J. Voller, S. Harrison (eds), Conservation biology principles for forested landscapes. Vancouver, UBC, pp. 98–129.

475 riparian forest

At a large scale, it is the band of forest that has a significant influence on a stream ecosystem or is significantly affected by the stream. At a smaller scale, it is the forest at the immediate water's edge, where some specialized plants and animals form a distinct community.

Source: Natural Resources Canada (eds), 1995. The State of Canada's forests 1994. Ottawa, Natural Resources Canada. Pp. 111.

492 riparian

Of vegetation growing in close proximity to a watercourse, lake, swamp or spring, and often dependent on its roots reaching the water table.
Trees forming a strip along a watercourse may be termed gallery forest = fringing forest.

Source: Ford-Robertson F.C. (ed), 1983. Terminology of forest science, technology, practice and products. 2nd printing with addendum. Washington, D.C., Society of American Foresters. Pp. 370.

450 Riparian Management Area (RMA)

A classified area of specified width surrounding or adjacent to streams, lakes, riparian areas, and wetlands. The RMA includes, in many cases, adjacent upland areas. It extends from the top of the streambank (bank full height) or from the edge of a riparian area or wetland or the natural boundary of a lake outward to the greater of: 1) the specified RMA distance, 2) the top of the inner gorge, or 3) the edge of the flood plain. Where a riparian area or wetland occurs adjacent to a stream or lake, the RMA is measured from the outer edge of the wetland.

Source: British Columbia Forest Service, March 1997: Glossary of Forestry Terms. Available from the World Wide Web: http://www.for.gov.bc.ca/pab/publctns/glossary/GLOSSARY.HTM. Cited 01-Oct-98.

514 Auwald

Laubmischwaldgesellschaften im Überflutungs- bzw. Strömungsgebiet der Flüsse.

Source: Brünig E., Mayer H., 1980. Waldbauliche Terminologie. Wien, Universität für Bodenkultur. Pp. 207. http://efern.boku.ac.at/forex/wbterm/

499 Auwald

Auwälder sind edaphisch bedingte Dauergesellschaften im Überschwemmungsgebiet der Flüsse.
(...) Charakteristisch sind kontinuierliche und plötzliche Standortsveränderungen durch Anlandung und Erosion. Auwälder unterliegen deshalb einer sehr differenzierten Entwicklungsdynamik.

Source: Mayer H., 1984. Wälder Europas. Stuttgart, Gustav Fischer. Pp. 691.

569 Auenwald

Im Überschwemmungsbereich von Flüssen gelegener Wald mit feuchtigkeitsliebenden, nährstoffreichen Boden bevorzugenden und an den stark wechselnden Grundwasserspiegel angepaßten Pflanzenarten. Je nach Häufigkeit und Dauer der Überschwemmung bilden sich Weichholzauen oder Hartholzauen.

Source: Schütt P., Schuck H.J., Stimm B. (eds), 1992. Lexikon der Forstbotanik. Landsberg/Lech (Germany), ecomed Verlag. Pp. 581.

570 Hartholzaue

Im Spitzenhochwasserbereich des Mittel- und Unterlaufs von Flüssen gelegener Teil des Auenwaldes, der nur selten und dann kurzzeitig überflutet wird. Hartholzauen gleichen Edellaubwäldern und werden deshalb in der pflanzensoziologischen Systematik im Verband der Erlen- und Edellaub-Auenwälder (Alno-Ulmion) zusammengefaßt.
Typische Vertreter der Hartholzauen sind Fraxinus exelsior, Prunus padus, Ulmus laevis und Ulmus minor, Quercus robur, die Kletterpflanzen Vitis vinifera, Humulus lupulus und Clematis vitalba, Nährstoffzeiger wie Sambucus nigra und Urtica dioica sowie Kryptophyten wie Ranunculus ficaria, Anemone ranunculoides, Galanthus nivalis und Arum maculatum.

Source: Schütt P., Schuck H.J., Stimm B. (eds), 1992. Lexikon der Forstbotanik. Landsberg/Lech (Germany), ecomed Verlag. Pp. 581.

571 Weichholzaue

Im Hochwasserbereich von Flüssen gelegener Teil des Auenwaldes, der von raschwüchsigen Bäumen mit relativ weichem, wenig haltbarem Holz besiedelt ist.

Source: Schütt P., Schuck H.J., Stimm B. (eds), 1992. Lexikon der Forstbotanik. Landsberg/Lech (Germany), ecomed Verlag. Pp. 581.

63 ripicole

(adj.) Se dit d'espèces, de communautés ou de formations végétales localisées au bord des cours d'eau.

Source: Delpech R., Dume G., Galmiche P., Timbal J., 1985. Vocabulaire: typologie des stations forestières. Paris, Institut pour le Développement Forestier. Pp. 243.

64 ripisylve

A éviter d'après Delpech et al. (1985). = forêt ripicole.

Source: Delpech R., Dume G., Galmiche P., Timbal J., 1985. Vocabulaire: typologie des stations forestières. Paris,

Institut pour le Développement Forestier. Pp. 243.

89 secondary forest

408 secondary forest
Forest that has regenerated after the original (primary) forest cover has been removed.

Source: Montreal Process Implementation Group, 1998. A framework of regional (sub-national) level criteria and indicators of sustainable forest management in Australia. Canberra, Commonweath of Australia. Pp. 108.

409 secondary forest
Natural forest growth after some major disturbance (e.g. logging, serious fire, or insect attack).

Source: Heywood V.H., Watson R.T., Baste I. (eds), 1995. Global biodiversity assessment. Cambridge, Cambridge University Press. Pp. 1140.

90 sensitive area

454 sensitive areas
Small areas designated to protect important values during forest and range operations. These areas, established by a Ministry of Forests district manager in consultation with a designated B.C. Environment official, guide operations on a site-specific basis and require a combination of forest practices. Sensitive areas will be mapped by resource agencies, and include regionally significant recreational areas, scenic areas with high visual quality objectives, and forest ecosystem networks.

Source: British Columbia Forest Service, March 1997: Glossary of Forestry Terms. Available from the World Wide Web: http://www.for.gov.bc.ca/pab/publctns/glossary/GLOSSARY.HTM. Cited 01-Oct-98.

455 sensitive slopes
Any slope identified as prone to mass wasting.

Source: British Columbia Forest Service, March 1997: Glossary of Forestry Terms. Available from the World Wide Web: http://www.for.gov.bc.ca/pab/publctns/glossary/GLOSSARY.HTM. Cited 01-Oct-98.

456 sensitive soils
Forest land areas that have a moderate to very high hazard for soil compaction, erosion, displacement, mass wasting or forest floor displacement.

Source: British Columbia Forest Service, March 1997: Glossary of Forestry Terms. Available from the World Wide Web: http://www.for.gov.bc.ca/pab/publctns/glossary/GLOSSARY.HTM. Cited 01-Oct-98.

457 sensitive watershed
A watershed that is used for domestic purposes or that has significant downstream fisheries values, and in which the quality of the water resource is highly responsive to changes in the environment. Typically, such watersheds lack settlement ponds, are relatively small, are located on steep slopes, and have special concerns such as extreme risk of erosion.

Source: British Columbia Forest Service, March 1997: Glossary of Forestry Terms. Available from the World Wide Web: http://www.for.gov.bc.ca/pab/publctns/glossary/GLOSSARY.HTM. Cited 01-Oct-98.

91 site

493 site
An area considered in terms of its environment, particularly as this determines the type and quality of the vegetation the area can carry.

Sites are classified either qualitatively, by their climate, soil and vegetation, into site types, or qualitatively, by their potential wood production, into site classes.

Source: Ford-Robertson F.C. (ed), 1983. Terminology of forest science, technology, practice and products. 2nd printing with addendum. Washington, D.C., Society of American Foresters. Pp. 370.

488 Standort
Gesamte Umwelt, die auf eine Pflanzengesellschaft einwirkt (Klima, Boden, Relief, andere Lebenswesen).

Source: Schütz J.P., Brang P., Bonfils P., Bucher H.U., 1993. Darstellung der Standortansprüche wichtiger Baumarten im Ökogramm und Gesellschaftsanschluss der Baumarten. In: Schmider P., Küper M., Tschander B., Käser B. (eds), Die Waldstandorte im Kanton Zürich. Vdf, Zürich, pp. 254–258.

549 Standort
(1) Ein bestimmter Teil der Erdoberfläche, der durch relativ einheitliche Verhältnisse und geographische Lage ausgezeichnet und abgegrenzt ist.

(2) Die reale und typische ökologische Umwelt (habitat) eines realen oder typischen Organismus (Einzelbaum) eines Taxons (Art, Rasse) oder einer heterogenen Gemeinschaft (Bestand, Pflanzengesellschaft).

(3) Die Geländequalität ohne Bezug auf die vorhandenen Organismen, z.B. die Gesamtheit der wachstumsbestimmenden, lage- und raumbezogenen Faktoren eines Geländeabschnittes.

Die räumliche Aneinanderreihung von Standorten ergibt die Standortsreihe (bodenkundlich Catena).

(a) Forstlicher Standort (forest site). Der Komplex physikalischer (klimatischer und edaphischer) und biologischer Faktoren (natürliche Umwelteinflüsse) bestimmt, welche Art von Wäldern und Beständen der Standort tragen kann. Die Eigenart des Standortes begrenzt die Wahl der Baumarten, des Bestandesaufbaus und der Betriebszieltypen und entscheidet über die potentielle Produktivität der Baumarten, Bestandestypen und Betriebsformen.

(b) Standortsklasse (site class). Die ertragskundliche Einstufung eines forstlichen Standortes hinsichtlich seines Ertragsvermögens an Forstprodukten und Dienstleistungen. Das Ertragsvermögen eines Standortes kann anhand der Wuchsleistung vorhandener Bestände oder durch Vergleich mit den Leistungen auf anderen Standorten mit analogen klimatischen und edaphischen Verhältnissen geschätzt werden.

Source: Brünig E., Mayer H., 1980. Waldbauliche Terminologie. Wien, Universität für Bodenkultur. Pp. 207.
http://efern.boku.ac.at/forex/wbterm/

92 *site adaptedness

507 Standortstolerance
Grössere Anpassungsfähigkeit von Baumarten (Ökotypen) an die örtlichen Klima- und Bodenbedingungen, z.B. Universalrassen.

Source: Brünig E., Mayer H., 1980. Waldbauliche Terminologie. Wien, Universität für Bodenkultur. Pp. 207. http://efern.boku.ac.at/forex/wbterm/

93 site class

440 site class
A classification of forest land in terms of inherent capacity to grow crops of industrial wood. The class identifies the average potential growth in cubic feet/acre/year (trees 5 inches diameter or larger to a 4-inch top) and is based on the culmination of mean annual increment of fully stocked natural stands.

Source: Southern Region Forest Inventory and Analysis, Forest Inventory – Definitions of Terms (no date). Available from the World Wide Web: http://www.srsfia.usfs.msstate.edu/fidef2.htm. Cited 01-Oct-98.

495 site class
A measure of the relative productive capacity of a site for the crop or stand under study, based e.g. on volume or height (dominant, co-dominant or mean) or the maximum mean annual increment, that is attained or attainable at a given age.

Source: Ford-Robertson F.C. (ed), 1983. Terminology of forest science, technology, practice and products. 2nd printing with addendum. Washington, D.C., Society of American Foresters. Pp. 370.

94 site type

494 site type
Sites are classified either qualitatively, by their climate, soil and vegetation, into site types, or qualitatively, by their potential wood production, into site classes.

Source: Ford-Robertson F.C. (ed), 1983. Terminology of forest science, technology, practice and products. 2nd printing with addendum. Washington, D.C., Society of American Foresters. Pp. 370.

504 Standortstyp
Zusammenfassung von waldbaulich-ökologisch gleichwertigen Standorten nach einheitlichen Merkmalen (Geologie, Klima, Boden) zu einem Typ für Zwecke der Kartierung, Planung und Behandlung (Standortseinheit).

Source: Brünig E., Mayer H., 1980. Waldbauliche Terminologie. Wien, Universität für Bodenkultur. Pp. 207.
http://efern.boku.ac.at/forex/wbterm/

95 species diversity

309 species diversity
One of the measures of biological diversity in forest ecosystems. The number of species in the ecosystem (species richness), or one of several indices that reflect the relative commonness of different species (species evenness).

Source: Kimmins J.P., 1997. Forest ecology. 2nd ed. Upper Saddle River, NJ, Prentice Hall. Pp. 596.

310 species diversity
We define and measure species diversity as a parameter of average rarity within a forest stand (Patil and Taillie 1982). In a diverse community, average rarity is the probability that a particular species will be comparatively rare; therefore, in a single community, such as a red pine (*Pinus resinosa* Ait.) plantation, the average rarity or diversity measure is expected to be near zero.

Source: Niese J.N., Strong T.F., 1992. Economic and tree diversity trade-offs in managed northern hardwoods. Canadian Journal of Forest Research 22: 1807–1813.

311 species diversity

Species diversity is a function of the number of species present (species richness or species abundance) and the evenness with which the individuals are distributed among these species (species evenness or species equitability) (Margalef 1958, Lloyd and Ghelardi 1964, Pielou 1966). If the term "species diversity" is to retain any usefulness (and this seems doubtful) its meaning probably should be restricted to at least this extent. Its use in other senses has been one cause of the term's present ambiguity. Some workers appear to synonymize species richness with species diversity or at least consider species richness to be one of several possible measures of species diversity (e.g. MacArthur 1965, Whittaker 1965, Paine 1966, Pianka 1966, 1967, Hutchinson 1967: 372, Hessler and Sanders 1967, MacArthur and Wilson 1967, Odum 1967, McNaughton 1967, 0968, Johnson, Mason , and Raven 1968, Sanders 1968, Whittaker and Woodwell 1969).

Source: Hurlbert S.H., 1971. The nonconcept of species diversity: a critique and alternative parameters. Ecology 52(4): 577–586.

96 species evenness

313 species evenness

Species evenness usually has been defined as the ratio of observed diversity to maximum diversity, the latter being said to occur when the species in a collection are equally abundant (...).

Source: Hurlbert S.H., 1971. The nonconcept of species diversity: a critique and alternative parameters. Ecology 52(4): 577–586.

97 species richness

312 species richness

Species richness can refer to the number of species present, without any particular regard for the exact area or number of individuals examined. However, it is useful to distinguish between numerical species richness (hereinafter referred to simply as species richness), the number of species present in a collection containing a specified number of individuals, or, possibly, amount of biomass; and areal species richness or species density (Simpson 1964), the number of species present in a given area or volume of the environment.

Source: Hurlbert S.H., 1971. The nonconcept of species

diversity: a critique and alternative parameters. Ecology 52(4): 577–586.

410 species richness

The number of species within a region. (A term commonly used as a measure of species diversity, but technically only one aspect of diversity).

Source: World Resources Institute (WRI), The World Conservation Union (IUCN), United Nations Environment Programme (UNEP), 1992. Global biodiversity strategy: guidelines for action to save, study, and use earth's biotic wealth sustainably and equitably. Washington, DC, World Resources Institute (WRI). Pp. 244.

98 species-area curve

314 species-area curve

A graphical representation of the rate of change in number of species in a sample plot as the size of the plot is increased.

Source: Kimmins J.P., 1997. Forest ecology. 2nd ed. Upper Saddle River, NJ, Prentice Hall. Pp. 596.

511 Arten-Arealkunde

Das Verhältnis von Artenzahl zur Flächengröße. Sie steigt mit zunehmender Flächengröße zuerst steil an und verläuft dann immer flacher.

Source: Brünig E., Mayer H., 1980. Waldbauliche Terminologie. Wien, Universität für Bodenkultur. Pp. 207.
http://efern.boku.ac.at/forex/wbterm/

99 stand

458 stand

A community of trees sufficiently uniform in species composition, age, arrangement, and condition to be distinguishable as a group from the forest or other growth on the adjoining area, and thus forming a silviculture or management entity.

Source: British Columbia Forest Service, March 1997: Glossary of Forestry Terms. Available from the World Wide Web:
http://www.for.gov.bc.ca/pab/publctns/glossary/GLOSSARY.HTM. Cited 01-Oct-98.

519 Bestand

Ein Kollektiv von in gegenseitiger Wechselwirkung stehenden Bäumen von ausreichender Einheitlichkeit nach Artenzusammensetzung, Entwicklungszustand, Alter, Struktur und Aufbau, um sie von anderen Beständen zu unterscheiden und von ausreichender Ausdehnung, um ein typisches Innenklima zu entwickeln. Der Bestand ist vielfach die kleinste Einheit für die Planung und Durchführung forstlicher, insbes. waldbaulicher Maßnahmen. Mindestfläche 1.0 (0.5) ha.

(1) Ausscheidender Bestand (thinnings): Der Teil des Bestandes. der zur Entnahme im Zuge der Durchforstung heransteht, bzw. natürlich abstirbt.

(2) Verbleibender Bestand (residual stand, remaining stand): Der nach erfolgter Durchforstung verbleibende Teil des Bestandes.

(3) Wildbiologisch ein Kollektiv von Wildtieren einer Art, (Stock), vgl. Wildbestand.

Source: Brünig E., Mayer H., 1980. Waldbauliche Terminologie. Wien, Universität für Bodenkultur. Pp. 207. http://efern.boku.ac.at/forex/wbterm/

574 Bestand
Baumkollektiv, das sich von der Umgebung durch Baumartenzusammensetzung, Bestandesalter oder Aufbau wesentlich unterscheidet (nach SAFE), im LFI mit einer Minimalfläche von 5 Aren.

Source: Brassel P., Brändli U.B. (Red.), 1999. Schweizerisches Landesforstinventar. Ergebnisse der Zweitaufnahme 1993–1995. Birmensdorf, WSL; Bern, BUWAL; Bern, Haupt. Pp. 442.

100 stand structure

459 stand structure
The distribution of trees in a stand, which can be described by species, vertical or horizontal spatial patterns, size of trees or tree parts, age, or a combination of these.

Source: British Columbia Forest Service, March 1997: Glossary of Forestry Terms. Available from the World Wide Web:
http://www.for.gov.bc.ca/pab/publctns/glossary/GLOSSARY.HTM. Cited 01-Oct-98.

101 stand type

460 stand types
See "stand", "stand structure".

Source: British Columbia Forest Service, March 1997: Glossary of Forestry Terms. Available from the World Wide Web:
http://www.for.gov.bc.ca/pab/publctns/glossary/GLOSSARY.HTM. Cited 01-Oct-98.

521 Bestandestyp
Bestände gleicher oder sehr ähnlicher Bestockung hinsichtlich Baumartenzusammensetzung (Holzsortenanteil), Struktur, Altersaufbau und Wuchsverhältnisse, die waldbaulich ähnlich behandelt werden können. Der Bestandeszieltyp ist der angestrebte Bestandeszustand (vgl. Betriebszieltyp).

Source: Brünig E., Mayer H., 1980. Waldbauliche Terminologie. Wien, Universität für Bodenkultur. Pp. 207. http://efern.boku.ac.at/forex/wbterm/

102 standard

411 standard
A standard is a set of principles, criteria and indicators, or at least some combinations of these hierarchical levels, that serves as a tool to promote sustainable forest management, as a basis for monitoring and reporting or as a reference for assessment of actual forest management.

The term "standard" is also used as reference for one particular aspect of forest management, e.g. desirable species composition, tolerable erosion levels etc. In [van Bueren and Blom 1997] the term "norm" is used for reference to one particular aspect.

Source: van Bueren E.M.L., Blom E.M., 1997. Hierarchical framework for the formulation of sustainable forest management standards. Leiden, The Netherlands, Backhuys Publishers. Pp. 82.

103 stressor indicator

412 stressor indicator
A characteristic of the environment that is suspected to elicit a change in the state of an ecological resource, and they include both natural and human-induced stressors. Selected stressor indicators will be monitored in EMAP only when a relationship between specific condition and stressor indicators are known or if a testable hypothesis can be formulated.

(New term 1993; replaces environmental indicator.)

Source: Environmental Monitoring and Assessment Program (EMAP). Master Glossary [online]. Updated 09/15/97 [cited 1998-08-31]. Available from the World Wide Web:
<http://www.epa.gov/emfjulte/html/glossary.html>

104 structural diversity

317 structural diversity
One of the measures of biological diversity in forest ecosystems. It refers to the variation in tree size and canopy layer-

ing, the variety of different life forms of vegetation (trees, herbs, shrubs, mosses, climbers, epiphytes, etc.), and the relative size and abundance of standing dead trees (snags) and decaying logs on the ground (coarse woody debris). Structural diversity refers to these features within a particular local ecosystem (alpha structural diversity), or to variations in them between local ecosystems across the local landscape (beta structural diversity).

Source: Kimmins J.P., 1997. Forest ecology. 2nd ed. Upper Saddle River, NJ, Prentice Hall. Pp. 596.

105 succession

231 allogenic succession
The changes in the living community and in the soil and microclimatic characteristics of an ecosystem as a result of alterations in the physical environment that are independent of changes in the living community. Floods, wind, landslides, fire, drought, climate change, or sediment deposition by a river are examples of factors that can cause allogenic succession.

Source: Kimmins J.P., 1997. Forest ecology. 2nd ed. Upper Saddle River, NJ, Prentice Hall. Pp. 596.

249 autogenic succession
The change in the living community and in the soil and microclimatic characteristics of an ecosystem caused by the living community itself (auto=self; genic=caused or generated); refers to effects caused by the plant community. The major mechanisms of autogenic succession are invasion and colonization, environmental alteration, and species exclusion.

Source: Kimmins J.P., 1997. Forest ecology. 2nd ed. Upper Saddle River, NJ, Prentice Hall. Pp. 596.

251 biogenic succession
The change in the living community and in the soil and microclimatic characteristics of an ecosystem caused mainly by the animal and microbial components of the living community. Insect outbreaks, mammalian herbivores, and epidemics of plant disease organisms are examples of biotic factors that can substantially alter the patterns of ecosystem development that would result from autogenic or allogenic processes acting alone.

Source: Kimmins J.P., 1997. Forest ecology. 2nd ed. Upper Saddle River, NJ, Prentice Hall. Pp. 596.

274 ecological succession
The process by which a series of different plant communities and associated animals and microbes successively occupy and replace each other over time in a particular eco-

system or landscape location following a disturbance to that ecosystem. Includes the accompanying change in the nonliving environment (soil and microclimate).

Source: Kimmins J.P., 1997. Forest ecology. 2nd ed. Upper Saddle River, NJ, Prentice Hall. Pp. 596.

512 succession
The non-seasonal, directional and continuous pattern of colonization and extinction on a site by populations.

Source: Begon M., Harper J.L., Townsend C.R., 1996. Ecology: Individuals, populations and communities. Third edition. Oxford, Blackwell. Pp. 1068.

553 allogenetische Sukzession
Der Wechsel wird durch Änderung der Standortsverhältnisse von außen verursacht (z.B. forstliche Maßnahmen).

Source: Brünig E., Mayer H., 1980. Waldbauliche Terminologie. Wien, Universität für Bodenkultur. Pp. 207.
http://efern.boku.ac.at/forex/wbterm/

554 autogenetische Sukzession
Die Pflanzengesellschaft verändert ihre eigene Umwelt (habitat) in einer Weise, daß sie von einer anderen Gesellschaft verdrängt werden kann (z.B. autotoxische Substanzen, Humusanhäufungen).

Source: Brünig E., Mayer H., 1980. Waldbauliche Terminologie. Wien, Universität für Bodenkultur. Pp. 207.
http://efern.boku.ac.at/forex/wbterm/

555 primäre Sukzession
Beginnt auf vegetationslosem Substrat (Lava, Küstenschlick, Geröll, Sanddüne, verlandendes Seeufer) und führt durch ein Kontinuum von Serien oder Phasen zur Schlußgesellschaft. Auch nach natürlichen Katastrophen (Feuer, Insekten, Überflutung).

Source: Brünig E., Mayer H., 1980. Waldbauliche Terminologie. Wien, Universität für Bodenkultur. Pp. 207.
http://efern.boku.ac.at/forex/wbterm/

556 sekundäre Sukzession
Beginnt nach mehr oder weniger vollständiger Zerstörung der ursprünglichen Vegetation infolge direkter oder indirekter Einflußnahme durch den Menschen und führt durch eine Folge von Gesellschaftstypen als Subserie schließlich zurück zum ursprünglichen Klimax, oder bleibt bei andauernder Einwirkung des auslösenden oder anderer Umweltfaktoren (z.B. Feuer) auf einem Stadium

des Disklimax oder Plagioklimax fixiert (z.B. Savannen). (Sekundärwald).

Source: Brünig E., Mayer H., 1980. Waldbauliche Terminologie. Wien, Universität für Bodenkultur. Pp. 207.
http://efern.boku.ac.at/forex/wbterm/

491 Sukzession
Natürliche Abfolge von Entwicklungsphasen im Wald, z.B. Kahlfläche – Pionierwald – Optimalphase der Baumentwicklung – Alters- und Zerfallsphase – wieder Kahlfläche oder Hochstaudenflur – Pionierwald usw.

Source: Schütz J.P., Brang P., Bonfils P., Bucher H.U., 1993. Darstellung der Standortansprüche wichtiger Baumarten im Ökogramm und Gesellschaftsanschluss der Baumarten. In: Schmider P., Küper M., Tschander B., Käser B. (eds), Die Waldstandorte im Kanton Zürich. Vdf, Zürich, pp. 254–258.

552 Sukzession
Eine zeitliche Folge von Veränderungen der Struktur und Funktionen der Vegetation (und ihrem Standort) infolge unterschiedlicher Wachstums- und Regenerationsraten sowie Konkurrenzwirkungen der sie zusammensetzenden Pflanzenarten.

(1) Das Verdrängen einer Pflanzengesellschaft durch eine andere im Verlauf ontogenetischer oder phylogenetischer Zeiträume.

(2) Der fortschreitende Wechsel der Vegetation von der Pioniergesellschaft bis zur Schlußgesellschaft.

Allogenetische (allogene) Sukzession (allogenic succession). Der Wechsel wird durch Änderung der Standortsverhältnisse von außen verursacht (z.B. forstliche Maßnahmen).

Autogenetische (autogene) Sukzession (autogenic succession). Die Pflanzengesellschaft verändert ihre eigene Umwelt (habitat) in einer Weise, daß sie von einer anderen Gesellschaft verdrängt werden kann (z.B. autotoxische Substanzen, Humusanhäufungen).

Primäre Sukzession (primary succession). Beginnt auf vegetationslosem Substrat (Lava, Küstenschlick, Geröll, Sanddüne, verlandendes Seeufer) und führt durch ein Kontinuum von Serien oder Phasen zur Schlußgesellschaft. Auch nach natürlichen Katastrophen (Feuer, Insekten, Überflutung).

Sekundäre Sukzession (secondary succession). Beginnt nach mehr oder weniger vollständiger Zerstörung der ursprünglichen Vegetation infolge direkter oder indirekter Einflußnahme durch den Menschen und führt durch eine Folge von Gesellschaftstypen als Subserie schließlich zurück zum ursprünglichen Klimax, oder bleibt bei andauernder Einwirkung des auslösenden oder anderer Umweltfaktoren (z.B. Feuer) auf einem Stadium des Disklimax oder Plagioklimax fixiert (z.B. Savannen).

(Sekundärwald).

Verschiedenen Phasen (vgl. Phase) einer Waldsukzessionsreihe werden gekennzeichnet durch bestimmte physio-ökologische Eigenschaften der Baumarten.

a) Pionierbaumarten (pioneers, ephemerals), kurzlebig (10–20 Jahre), Weichholz, breitkronig, lichtbedürftig, rasch wachsend;

b) Hauptphasenbaumarten (persistent, serals), länger lebend (etwa 50–300 Jahre), in der Jugend schnell wachsend, dann langandauerndes langsameres Wachstum, Licht- bis Schattbaumarten;

c) Schlußphasenbaumarten (slow-living, late phase species), langlebige, langsam wachsende, dichtkronige, schattenertragende Baumarten.

Eine genaue Kenntnis der natürlichen Boden- und Vegetationsentwicklung mit deutlich unterscheidbaren Phasen (Stadien) ist Voraussetzung für gezielte waldbauliche Maßnahmen, z.B. Wiederbesiedlung von Blaiken. Aus dem räumlichen Nebeneinander von Sukzessionstadien darf nicht ohne weiteres auf ein zeitliches Nacheinander geschlossen werden.

Source: Brünig E., Mayer H., 1980. Waldbauliche Terminologie. Wien, Universität für Bodenkultur. Pp. 207.
http://efern.boku.ac.at/forex/wbterm/

106 sustainable forest management

413 sustainable forest management
Sustainable forest management is the process of managing permanent forest land to achieve one or more clearly specified objectives of management with regard to the production of a continuous flow of desired forest products and service without undue reduction of its inherent values and future productivity and without undue undesirable effects on the physical and social environment. (Definition adopted by the International Tropical Timber Council (ITTC) in 1992.)

Source: van Bueren E.M.L., Blom E.M., 1997. Hierarchical framework for the formulation of sustainable forest management standards. Leiden, The Netherlands, Backhuys Publishers. Pp. 82.

414 sustainable forest management
Sustainable forest management means the stewardship and use of forests and forest lands in a way, and a rate, that maintains their biological diversity, productivity, regeneration capacity, vitality and their potential to fulfil, now and in the future, relevant ecological, economic and social functions, at local, national and global levels, and that does not cause damage to other ecosystems. (Ministerial Conference on the Protection of European Forests, 1993. Resolution H1.)

Source: Granholm H., Vähänen T., Sahlberg S. (compilers), 1996. Background document. Intergovernmental Seminar on Criteria and Indicators for Sustainable Forest Management, August 19–22, 1996, Helsinki. Helsinki, Ministry of Agriculture and Forestry, Intergovernmental Seminar on Criteria and Indicators for Sustainable Forest Management. Pp. 131.

107 temporal diversity

320 temporal diversity
The change over time in measures of alpha species and structural diversity, or in measures of landscape (beta) diversity, due to disturbance and autogenic succession.

Source: Kimmins J.P., 1997. Forest ecology. 2nd ed. Upper Saddle River, NJ, Prentice Hall. Pp. 596.

108 texture

515 texture measures
This is NOT a definition!
Texture measures are used to anlayse patterns of brigthness variation within an image (...). These measures can be used profitably in landscape ecology to analyse the complexity of the mosaic and the contrast between patches.

Source: Farina A., 1998. Principles and methods in landscape ecology. London, Chapman and Hall. Pp. 235.

109 threatened species

415 threatened species
A general term to denote species which are endangered, vulnerable, rare, indeterminate, or insufficiently known.

Source: UCN Conservation Monitoring Centre, Cambridge U.K. (compiler), 1994. IUCN red list of threatened animals. Gland, IUCN.

416 threatened species
Species that are, often genetically impoverished, of low fecundity, dependent on patchy or unpredictable resources, extremely variable in population density, persecuted or otherwise prone to extinction in human-dominated landscapes.

Source: Heywood V.H., Watson R.T., Baste I. (eds), 1995. Global biodiversity assessment. Cambridge, Cambridge University Press. Pp. 1140.

110 tree

436 tree
Woody plants that generally have a single main stem and have more or less definite crowns. In instances where life form cannot be determined, woody plants equal to or greater than 5 m in height will be considered trees.

Source: Vegetation Subcommittee, Federal Geographic Data Committee, June 1997: Vegetation Classification Standard. Appendix III: Definitions (normative). Available on World Wide Web: http://biology.usgs.gov/fgdc.veg/standards/appendix3.htm Last updated: 06-Nov-1997. Cited: 01-Oct-98.
http://biology.usgs.gov/fgdc.veg/standards/appendix3.htm

441 tree
A woody plant usually having one or more perennial stems, a more or less definitely formed crown of foliage, and a height of al least 12 feet at maturity.

Source: Southern Region Forest Inventory and Analysis, Forest Inventory – Definitions of Terms (no date). Available from the World Wide Web: http://www.srsfia.usfs.msstate.edu/fidef2.htm. Cited 01-Oct-98.

442 tree
A woody, self-supporting perennial plant usually with a single main stem and generally growing more than 20 feet tall.

Source: GardenWeb, 1998: Glossary of Botanical Terms. Available from the World Wide Web: http://www.gardenweb.com/glossary/ Last updated 26-Sept-98. Cited 01-Oct-98.

477 tree
A generic term for tree species of any size – seedlings, saplings, subplot trees, timberland species, and woodland species. Trees less than 5.0 inches (12.7 cm) DBH/DRC are measured on microplots, larger trees are measured on subplots.

Source: USDA Forest Service, 1998. Forest health monitoring 1998. Field methods guide. Research Triangle Park, NC, USDA Forest Service, National Forest Health Monitoring Program. Pp. [mult. pag.].

534 tree
A woody perennial with a single main stem or, in the case of coppice, with several stems, having a more or less definite crown.
   Includes: bamboos, palms and other woody plants meeting the above criterion.

Source: UN-ECE/FAO , 1997. UN-ECE/FAO Temperate and Boreal Forest Resources Assessment 2000. Terms and definitions. New York and Geneva, United Nations. Pp. 13.

515 Baum
Eine perennierende Pflanze mit einem durchgehenden, verholzten Stamm und einer mehr oder weniger wohlausgebildeten Krone; in der Regel über 5 m hoch werdend.

Source: Brünig E., Mayer H., 1980. Waldbauliche Terminologie. Wien, Universität für Bodenkultur. Pp. 207.
http://efern.boku.ac.at/forex/wbterm/

111 umbrella species

417 umbrella species
Species whose occupancy area (plants) or home range (animals) are large enough and whose habitat requirements are wide enought that, if they are given a sufficiently large area for their protection, will bring other species under that protection.

Source: Heywood V.H., Watson R.T., Baste I. (eds), 1995. Global biodiversity assessment. Cambridge, Cambridge University Press. Pp. 1140.

224 umbrella species
A large-bodied, popular species having a large home range and broad requirements for habitats and resources, that can be managed to also provide habitats and resources for other species; similar to flagship species.

Source: Marcot B.G., Wisdom M.H., Li H.W., Castillo G.C., 1994. Managing for featured, threatened, endangered, and sensitive species and unique habitats for ecosystem sustainability. General Technical Report PNW 329, Portland, OR, USDA Forest Service, Pacific Northwest Research Station, 39 p.

112 unitype species

225 unitype species
A wildlife species that uses and requires only one kind of habitat or successional stage, typically their interiors.

Source: Marcot B.G., Wisdom M.H., Li H.W., Castillo G.C., 1994. Managing for featured, threatened, endangered, and sensitive species and unique habitats for ecosystem sustainability. General Technical Report PNW 329, Portland, OR, USDA Forest Service, Pacific Northwest Research Station, 39 p.

113 *unnatural

483 naturfremder Bestand
Bestand mit hohem Anteil an standortfremden Baumarten.

Source: Schütz J.P., Brang P., Bonfils P., Bucher H.U., 1993. Darstellung der Standortansprüche wichtiger Baumarten im Ökogramm und Gesellschaftsanschluss der Baumarten. In: Schmider P., Küper M., Tschander B., Käser B. (eds), Die Waldstandorte im Kanton Zürich. Vdf, Zürich, pp. 254–258.

114 *unsuitable species

503 standortwidrig
Die spezifische Baumart eignet sich nicht für den Anbau am betreffenden Standort, da Wuchs, Schädlingsresistenz, Ausformung und nachhaltige Massen- und Wertleistung nicht befriedigen.

Source: Brünig E., Mayer H., 1980. Waldbauliche Terminologie. Wien, Universität für Bodenkultur. Pp. 207.
http://efern.boku.ac.at/forex/wbterm/

587 standortswidrig
Bei Vorkommen den Standort schädigend.
   Auch die Schreibweise ohne Genitiv-S im Wortmitte kommt vor.

Source: P. Brang (WSL Birmensdorf).

115 vegetation

437 vegetation
The collective plant cover over an area.

Source: Vegetation Subcommittee, Federal Geographic Data Committee, June 1997: Vegetation Classification Standard. Appendix III: Definitions (normative). Available on World Wide Web: http://biology.usgs.gov/fgdc.veg/standards/appendix3.htm. Last updated: 06-Nov-1997. Cited: 01-Oct-98.
http://biology.usgs.gov/fgdc.veg/standards/appendix3.htm

560 Vegetation
(1) Pflanzen allgemein und die Gesamtheit des Pflanzenlebens in einem Gebiet;
   (2) Potentiell natürliche Vegetation: Pflanzengesellschaft, die sich heute auf Grund der Umweltbedingungen eines Standortes vorfindet oder langfristig wieder einstellt, wenn der Einfluß des Menschen ausgeschaltet wurde.

Source: Brünig E., Mayer H., 1980. Waldbauliche Terminologie. Wien, Universität für Bodenkultur. Pp. 207.
http://efern.boku.ac.at/forex/wbterm/

572 Vegetation
Gesamtheit der Pflanzengesellschaften eines Gebietes.

Source: Schütt P., Schuck H.J., Stimm B. (eds), 1992. Lexikon der Forstbotanik. Landsberg/Lech (Germany), ecomed Verlag. Pp. 581.

116 *vegetation type

561 Vegetationstyp
Eine bestimmte Form der Vegetation oder Pflanzengesellschaft gleich welcher Größenordnung (Flächenausdehnung), Rang oder Phase der Sukzession.

Source: Brünig E., Mayer H., 1980. Waldbauliche Terminologie. Wien, Universität für Bodenkultur. Pp. 207.
http://efern.boku.ac.at/forex/wbterm/

117 verifier

418 verifier
A verifier is the source of information for the indicator or for the reference value for the indicator.

Source: van Bueren E.M.L., Blom E.M., 1997. Hierarchical framework for the formulation of sustainable forest management standards. Leiden, The Netherlands, Backhuys Publishers. Pp. 82.

118 vulnerable species

419 vulnerable species
10% probability of extinction within 100 years.

Source: Mace G.M., Lande R., 1995. Assessing extinction threats: toward a reevaluation of IUCN threatened species categories. In: Ehrenfeld D. (ed), To preserve biodiversity: an overview. Cambridge, Massachusetts, Blackwell Science, pp. 51–60.

420 vulnerable species
Taxa believed likely to move into the "Endangered" category in the near future if the causal factors continue operating. Included are taxa of which most or all the populations are decreasing because of over-exploitation, extensive destruction of habitat or other environmental disturbance. Taxa with populations that are still abundant but are under threat from severe adverse factors throughout their range [sic]. In practice, "Endangered" and "Vulnerable" categories may include, temporarily, taxa whose populations are beginning to recover as a result of remedial action, but whose recovery is insufficient to justify their transfer to another category.

Source: UCN Conservation Monitoring Centre, Cambridge U.K. (compiler), 1994. IUCN red list of threatened animals. Gland, IUCN. Cited in Loiskekosti M., Halko L. (eds), 1993. Ministerial Conference on the Protection of Forests in Europe, 16–17 June 1993 in Helsinki. European list of criteria and most suitable quantitative indicators. Helsinki, Ministry of Agriculture and Forestry. Pp. 20.

References

Aird P.L., 1994. Conservation for the sustainable development of forests worldwide – a compendium of concepts and terms. Forestry Chronicle 70(6): 666–674.

Allen T.F.H., Starr T.B., 1982. Hierarchy. Perspectives for ecological complexity. Chicago, University of Chicago Press. 310 pp

Begon M., Harper J.L., Townsend C.R., 1996. Ecology: Individuals, populations and communities. Third edition. Oxford, Blackwell. 1068 pp.

Bibby C.J., Collar N.J., Crosby M.J., Heath M.F., Imboden C., Johnson T.H., Long A.J., Stattersfield A.J., Thirgood S.J., 1992. Putting biodiversity on the map: priority areas for global conservation. (2nd printing). Girton, International Council for Bird Preservation. 90 pp.

Bisby F.A., Coddington J., Thrope J.P., Smartt J., Hengeveld R., Edwards P.J., Duffield S.J. (lead authors), 1995. Characterization of biodiversity. In: Heywood V.H., Watson R.T., Baste I. (eds), Global biodiversity assessment. Cambridge, Cambridge University Press, pp. 21–106.

Brassel P., Brändli U.B. (Red.), 1999. Schweizerisches Landesforstinventar. Ergebnisse der Zweitaufnahme 1993–1995. Birmensdorf, WSL; Bern, BUWAL; Bern, Haupt. 442pp.

British Columbia Forest Service, March 1997: Glossary of Forestry Terms. Available from the World Wide Web: http://www.for.gov.bc.ca/pab/publctns/glossary/GLOSSARY.HTM. Cited 01-Oct-98.

Bruenig E.F., 1996. Conservation and management of tropical rainforests: an integrated approach to sustainability. Wallingford, CAB International. 339 pp.

Brünig E., Mayer H., 1980. Waldbauliche Terminologie. Wien, Universität für Bodenkultur. Pp. 207. Also available from the World Wide Web: <http://efern.boku.ac.at/forex/wbterm/>

Bundesministerium für Ernährung, Landwirtschaft und Forsten: Bundesinventur 1986–1990, Vol. I, p. 115. Translation and comments: J. Innes, H. Ellenberg, Jyllinge. Cited in: Frank G., Halbritter K., 1998: Regional BEAR support meeting, 11–12 July 1998, Göttingen, Germany. Minutes. 15 pp.

Chauvet M., Olivier L., 1993. La biodiversité, enjeu planétaire: préserver notre patrimoine génétique. Paris, Sang de la Terre. 413 pp.

Colinvaux P.A., 1993. Ecology 2. 2nd ed. New York, Wiley. 688 pp.

Criteria and indicators for the conservation and sustainable management of temperate and boreal forests. The Montreal Process, December 1998, 15 p.Dasmann R.F., 1991. The importance of cultural and biological diversity. In: Oldfield M.L., Alcorn J.B. (eds), Biodiversity: culture, conservation, and ecodevelopment. Boulder, Westview Press, pp. 7–15.

Delpech R., Dume G., Galmiche P., Timbal J., 1985. Vocabulaire: typologie des stations forestières. Paris, Institut pour le Développement Forestier. 243 pp.

di Castri F., Younès T., 1995. Introduction: Biodiversity, the emergence of a new scientific field – its perspectives and constraints. In: di Castri F., Younès T. (eds), Biodiversity, science and development: towards a new partnership. Wallingford, CAB International, pp. 1–11.

Dudley N., 1996. Authenticity as a means of measuring forest quality. Biodiversity Letters 3(1): 6–9.

Dunster K., Dunster J., 1996. Dictionary of natural resource management. Vancouver, UBC Press. 363 pp.

Eldredge N., 1992. Where the twain meet: causal intersections between the genealogical and ecological realms. In: Eldredge N. (ed), Systematics, ecology, and the biodiversity crisis. New York, Columbia University Press, pp. 1–14.

Ellenberg H., Klötzli F., 1972. Waldgesellschaften und Waldstandorte der Schweiz. Mitteilungen EAFV 48: 587–930.

Environmental Monitoring and Assessment Program (EMAP). Master Glossary [online]. Updated 09/15/97 [cited 1998-08-31]. Available from World Wide Web: <http://www.epa.gov/emfjulte/html/glossary.html>

Farina A., 1998. Principles and methods in landscape ecology. London, Chapman and Hall. 235 pp.

Food and Agriculture Organization (ed), 1997. State of the world's forests 1997. Rome, Food and Agriculture Organization. Pp. 200. http://www.fao.org/waicent/faoinfo/forestry/SOFOTOC.htm

Ford-Robertson F.C. (ed), 1983. Terminology of forest science, technology, practice and products. 2nd printing with addendum. Washington, D.C., Society of American Foresters. 370 pp.

Forman R.T.T., 1995. Land mosaics. The ecology of landscapes and regions. Cambridge, Cambridge University Press. 632 pp.

Forman R.T.T., Godron M., 1986. Landscape ecology. New York, Wiley. 619 pp.

Freedman B., 1989. Environmental ecology. The impacts of pollution and other stresses on ecosystem structure and function. San Diego, Academic Press. 424 pp.

GardenWeb, 1998: Glossary of Botanical Terms. [Online] Available from the World Wide Web: http://www.gardenweb.com/glossary/ Last updated 26-Sept-98. Cited 01-Oct-98.

Gilpin M.E., Soulé M.E., 1986. Minimum viable populations: processes of species extinction. In: M.E. Soulé (ed), Conservation biology: the science of scarcity and diversity. Sunderland, Mass., Sinauer Associates, Inc.

Granholm H., Vähänen T., Sahlberg S. (compilers), 1996. Background document. Intergovernmental Seminar on Criteria and Indicators for Sustainable Forest Management, August 19–22, 1996, Helsinki. Helsinki, Ministry of Agriculture and Forestry, Intergovernmental Seminar on Criteria and Indicators for Sustainable Forest Management. 131 pp.

Haddon B.D. (ed), 1988. Forest inventory terms in Canada / Terminologie de l'inventaire des forêts du Canada. 3rd ed. Ottawa, Canadian Forest Inventory Committee, Forestry Canada. 113 + 109 pp.

Haila V., Kouki J., 1994. The phenomenon of biodiversity in conservation biology. Annales Zoologici Fennici 31(1): 5–18.

Hellawell J.M., 1991. Development of a rationale for monitoring. In: B. Goldsmith (ed), Monitoring for conservation and ecology. London, Chapman and Hall, pp. 1–14.

Helms J.A. (ed), 1998. The Dictionary of Forestry. Bethesda, MD, Society of American Foresters. 224 pp.

Heywood V.H., Baste I., Gardner K.A. (lead authors), 1995. Introduction. In: Heywood V.H., Watson R.T., Baste I. (eds), Global biodiversity assessment. Cambridge, Cambridge University Press, pp. 1–19.

Heywood V.H., Watson R.T., Baste I. (eds), 1995. Global biodiversity assessment. Cambridge, Cambridge University Press. 1140 pp.

Hurlbert S.H., 1971. The nonconcept of species diversity: a critique and alternative parameters. Ecology 52(4): 577–586.

Jensen E.C., Anderson D.J., 1995. The reproductive ecology of broadleaved trees and shrubs: glossary. Research Contribution 9f, Corvallis, Forest Research Laboratory, Oregon State University, 8 pp.

Kimmins J.P., 1997. Forest ecology. 2nd ed. Upper Saddle River, NJ, Prentice Hall. 596 pp.

Koch G., Kirchmeir H., 1997. Methodik der Hemerobiebewertung. Österreichische Forstzeitung (1): 24–26.

Köhl M., Päivinen R., 1996. Definition of a system of nomenclature for mapping European forests and for compiling a pan-European forest information system. Luxembourg, Office for Official Publications of the European Communities. 238 pp.

Kowarik I., 1988. Zum menschlichen Einfluß auf Flora und Vegetation. Theoretische Konzepte und ein Quantifizierungsansatz am Beispiel von Berlin (West). Landschaftsentwicklung und Umweltforschung 56, Berlin, Technische Universität Berlin, 280 S.

Loiskekosti M., Halko L. (eds), 1993. Ministerial Conference on the Protection of Forests in Europe, 16–17 June 1993 in Helsinki. European list of criteria and most suitable quantitative indicators. Helsinki, Ministry of Agriculture and Forestry. 20 pp.

Lund G. (ed), 1998. Definitions of deforestation, afforestation and reforestation [on line]. Cited 2-Oct-98. Available from the World Wide Web <http://home.att.net/~gklund/DEFpaper.html>

LWF/SSI (WSL Birmensdorf): Ad hoc group of experts for the standardisation of the terminology to be used in the Sanasilva report 1997 (Brang P. (Redaktion), 1998. Sanasilva-Bericht 1997. Berichte der Eidg. Forschungsanstalt für Wald, Schnee und Landschaft 345, 102 pp.).

Mace G.M., Lande R., 1995. Assessing extinction threats: toward a reevaluation of IUCN threatened species categories. In: Ehrenfeld D. (ed), To preserve biodiversity: an overview. Cambridge, Massachusetts, Blackwell Science, pp. 51–60.

Marcot B.G., Wisdom M.H., Li H.W., Castillo G.C., 1994. Managing for featured, threatened, endangered, and sensitive species and unique habitats for ecosystem sustainability. General Technical Report PNW 329, Portland, OR, USDA Forest Service, Pacific Northwest Research Station, 39 p.

Markert B., Oehlmann J., Roth M., 1997. Biomonitoring von Schwermetallen – eine kritische Bestandsaufnahme. Zeitschrift für Ökologie und Naturschutz 6: 1–8.

Mayer H., 1984. Wälder Europas. Stuttgart, Gustav Fischer. 691 pp.

Meffe G.K., Carroll C.R. (eds), 1994. Principles of conservation biology. Sunderland, Ma, Sinauer. 600 pp.

Montreal Process Implementation Group, 1998. A framework of regional (sub-national) level criteria and indicators of sustainable forest management in Australia. Canberra, Commonweath of Australia. 108 pp.

Natural Resources Canada (eds), 1995. The State of Canada's forests 1994. Ottawa, Natural Resources Canada. 111 pp.

Niese J.N., Strong T.F., 1992. Economic and tree diversity trade-offs in managed northern hardwoods. Canadian Journal of Forest Research 22: 1807–1813.

Panel of the Board on Science and Technology for International Development, 1992. Conserving biodiversity: a research agenda for development agencies. Report of a Panel of the Board on Science and Technology for International Development U.S. National Research Council. (2nd printing). Washington, D.C., National Academy Press. 127 pp.

Parminter J., 1998. Natural disturbance ecology. In: J. Voller, S. Harrison (eds), Conservation biology principles for forested landscapes. Vancouver, UBC, pp. 3–41.

Policy and Economics Directorate Forestry Canada / Direction Générale des politiques et de l'Economie, Forêts Canada, 1992. Silvicultural terms in Canada / Terminologie de la sylviculture au Canada (trad. et adapté de la version anglaise). Ottawa, Science and Sustainable Development Directorate, Forestry Canada. 74 pp.

Ramade F., 1993. Dictionnaire encyclopédique de l'écologie et des sciences de l'environnement. Paris, Ediscience international. 822 pp.

Reid W.V., Miller K.R., 1989. Keeping options alive: the scientific basis for conserving biodiversity. Washington, World Resources Institute. 128 pp.

Roberts M.R., Gilliam F.S., 1995. Patterns and mechanisms of plant diversity in forested ecosystems: implications for forest management. Ecological Applications 5(4): 969–977.

Sandlund O.T., Hindar K., Brown A. H. D. (eds), 1992. Conservation of biodiversity for sustainable development. Oslo, Scandinavian University Press. 324 pp.

Schaefer M., 1992. Wörterbuch der Biologie. Ökologie. 3. Auflage. Jena, Gustav Fischer Verlag. 433 pp.

Schütt P., Schuck H.J., Stimm B. (eds), 1992. Lexikon der Forstbotanik. Landsberg/Lech (Germany), ecomed Verlag. 581 pp.

Schütz J.P., Brang P., Bonfils P., Bucher H.U., 1993. Darstellung der Standortansprüche wichtiger Baumarten im Ökogramm und Gesellschaftsanschluss der Baumarten. In: Schmider P., Küper M., Tschander B., Käser B. (eds), Die Waldstandorte im Kanton Zürich. Vdf, Zürich, pp. 254–258.

Solbrig O.T., 1994. Biodiversity: an introduction. In: Solbrig O.T., van Emden H.M., van Oordt P.G.W.H.J. (eds), Biodiversity and global change. Wallingford, CAB International, International Union of Biological Sciences, pp. 13–20.

Southern Region Forest Inventory and Analysis, Forest Inventory – Definitions of Terms (no date). Available from the World Wide Web: http://www.srsfia.usfs.msstate.edu/fidef2.htm. Cited 01-Oct-98

Sukopp H., 1972. Wandel von Flora und Vegetation in Mitteleuropa unter dem Einfluss des Menschen. Berichte über Landwirtschaft 50: 112–139.

Tallent-Halsell N.G. (ed), 1994. Forest health monitoring 1994. Field methods guide. EPA/620/R-94/027. Washington, D.C., U.S. Environmental Protection Agency. [mult. pag.].

UCN Conservation Monitoring Centre, Cambridge U.K. (compiler), 1994. IUCN red list of threatened animals. Gland, IUCN.

UN-ECE/FAO, 1997. UN-ECE/FAO Temperate and Boreal Forest Resources Assessment 2000. Terms and definitions. New York and Geneva, United Nations. 13 pp.

United Nations Environment Programme (UNEP), 1992. Convention on Biological Diversity. Text and annexes. Geneva, UNEP. 34 pp.

USDA Forest Service, 1998. Forest health monitoring 1998. Field methods guide. Research Triangle Park, NC, USDA Forest Service, National Forest Health Monitoring Program. [mult. pag.].

van Bueren E.M.L., Blom E.M., 1997. Hierarchical framework for the formulation of sustainable forest management standards. Leiden, The Netherlands, Backhuys Publishers. 82 pp.

Vegetation Subcommittee, Federal Geographic Data Committee, June 1997: Vegetation Classification Standard. Appendix III: Definitions (normative). Available on World Wide Web: http://biology.usgs.gov/fgdc.veg/standards/appendix3.htm. Last updated: 06-Nov-1997. Cited: 01-Oct-98.

Voller J., 1998. Riparian areas and wetlands. In: J. Voller, S. Harrison (eds), Conservation biology principles for forested landscapes. Vancouver, UBC, pp. 98–129.

Wald und Boden. 1996. Schriftenreihe der Sächsischen Landesanstalt für Forsten 7/96.

Winkler S., 1980. Einführung in die Pflanzenökologie. Stuttgart, Gustav Fischer Verlag. 256 pp.

World Resources Institute (WRI), The World Conservation Union (IUCN), United Nations Environment Programme (UNEP), 1992. Global biodiversity strategy: guidelines for action to save, study, and use earth's biotic wealth sustainably and equitably. Washington, DC, World Resources Institute (WRI). 244 pp.

Appendix 7: BEAR Technical Report 1.

Indicators of biodiversity: recent approaches and some general suggestions

Background and early development

Biodiversity has been defined as all variation on the genetic, species and ecosystem levels, in agreement with the Rio Convention. Diversity on the landscape level ("gamma-diversity") may also have important effects on lower levels (e.g. Noss 1990, Hansson et al. 1995) and may be a determinant of large parts of regional diversity. It would be completely impossible to try to monitor all this potential variability and, indeed, only certain aspects of the biological variability may in the long run be important to retain: Invading species are usually considered undesirable. Local human disturbance can increase species richness. However, we still do not know which species are actually necessary for normal ecosystem functioning (Lawton 1994). Thus, there are two general problems in biodiversity management: 1) which is the important (or representative, or "valued") biodiversity for a certain system or region and 2) how in a fairly simple way make sure that that variability is retained.

The first problem has to be solved first. There has indeed been some recent progress in understanding the effects of varying biodiversity on ecosystem patterns and processes (e.g., Naeem et al. 1994, Grime 1997, Jones et al. 1997). The extinction risk of endangered herbivores has been related to the level of plant biodiversity (Ritchie 1999). And the ethical arguments for preserving biodiversity are stressed repeatedly.

The second problem is related to a selection of ecological indicators. An indicator may be a species, a structure, a process or some other feature of a biological system, the occurrence of which insures the maintenance or restoration of the most important aspects of biodiversity for that system. Diversity is often equated with species richness, although this is not in agreement with the Rio Convention. Some authors have instead stressed the importance of ecological mechanisms (Noss 1990) and keystone species (or "drivers" instead of "passengers" in the ecological systems, Walker 1992). The concept of "biotic integrity" (Anger-

maier and Karr 1994) covers biological diversity but also includes the ability of an ecosystem to function and maintain itself, including its native biodiversity. Certain indicator systems have already been developed with this biotic integrity particularly in mind.

The interest in indicators has a long history within ecology. The earliest use was probably to manually demarcate various plant associations within phytosociology. Such a function is still retained in more recent and advanced methodology in that discipline, e.g. in the TWINSPAN (Two-Way INdicator SPecies Analysis, however see Dufrene and Legendre 1997 for recent criticism) statistical program to separate various vegetation units in a tangled mosaic. Indicators have been common in ecotoxicology, e.g. as laboratory systems to demonstrate possible toxic effects of environmental contaminants. Certain fish species have often been used in this context, and the miners' canary is another noteworthy example. Lichens have for long been known to be severely negatively affected by pollutants in the air. This relationship has been exploited for field monitoring of pollution, using several lichen species as indicators (Hawksworth and Rose 1976, Skye 1979).

Indicators have also been used to demonstrate general population trends, e.g. the declining brown hare for wildlife generally in the European agricultural landscapes. Particularly in the US, one species has been selected as an indicator for a whole guild or even an ecosystem ("Management Indicator Species" as the bald eagle or the Florida panther, e.g. Severinghaus 1981, Verner 1984). However, this approach has met with limited success (e.g. DeGraaf and Chadwick 1984, Landres et al. 1988). Finally, indicators have already been used in conservation biology, as umbrella species (Launer and Murphy 1994, usually large species with wide areal requirements, presumed to also cover the requirements of other species, e.g. tigers) or flagships (Noss 1990, large appealing species attracting interest to their ecosystem, e.g. pandas). Problems with these various approaches have been discussed and evaluated by Landres et al. (1988).

Lennart Hansson (lennart.hansson@nvb.slu.se), Dept of Conservation Biology, Swedish Univ. of Agricultural Science, Box 7002, SE-750 07 Uppsala, Sweden.

Policy indicators

The subsequent discussion about possible indicators of biodiversity has developed in different directions. Some authors have argued for indicators for policy-making and others have tried to develop indicators for practical use in management and monitoring. At policy-making, indicators may be used to compare different localities, regions or countries regarding the biodiversity, or care of biodiversity. Such indicators can also be used to set priorities for land use and for conservation projects. Reid et al. (1983) presented a list of 22 such indicators for genetic diversity (also for domesticated species), species diversity and community diversity. These indicators consisted, e.g., of number or percent of species threatened by extinction, number of endemic species, number of species with decreasing populations, percentage of area in strictly protected reserves and present crop area related to that area thirty years earlier. Most of these statistics are fairly readily available from official sources. They are usually only applicable on a regional level. They can hardly be used to survey the development or recovery of a separate threatened ecosystem, reserve or forest stand. Such problems require management indicators and mainly that latter type of indicators will be treated below.

Criteria and indicators proposed in the "Helsinki Process" for protection of biodiversity of Europan forests (Loiskekoski et al. 1994) may be considered mainly as policy indicators. They evidently need to be supplemented with management indicators.

Single species vs community indicators

Some earlier use of particular indicator species as a monitoring device has met with severe criticism: in ecotoxicology toxic effects on laboratory specimens may not mean anything to natural populations due to compensatory survival (e.g. Cairns 1986), and neither to whole communities due to overwhelming competition or predation effects. Likewise, in conservation single species may simply not cover the vulnerability of any extensive system due to complex niche diversification. Different species may also be limited in different ways, e.g. by specific food resources, by predation or by social factors. There may even be negative correlations between abundances of indicator species and certain other species if there is strong interspecific competition between them. An example of the limitation of a separate species is the spotted owl that has got the rank of an indicator species but does not indicate all the needs of sympatric threatened amphibians (Harrison and Fahrig 1995).

A solution to the problems with single conservation indicator species appears to be a limited group of species bet-

ter covering the environmental variability of concern (Landres et al. 1988, Wilcove 1990). Birds as a group have been advocated as such wide-spectrum indicators (Järvinen and Väisänen 1979), containing residents, short distance and long-distance migrants, short- and long-lived species, granivores and insectivores, etc. Conventional diversity indices utilised in community ecology (e.g. species richness and the Shannon-Wiener or Simpson indices) were supposed to be useful for comparisons. However, even in such cases problems may arise: the species richness of birds associated with lakes and wetlands has generally increased but this increase is mainly due to eutrophication or pollution. Certain bird communities of Baltic islands demonstrated a decrease in Shannon-Wiener index when protected (Väisänen and Järvinen 1977); however, the actual reason was an disproportionate increase in the herring gull while all other species also increased but less so.

Statistical indicators

Much recent work has centred around statistical indicators, i.e. single species or species groups that are strongly correlated with total species richness or with species richness within certain taxa. This approach thus neglects important aspects in the original definition of biodiversity. Williams and Gaston (1994) proposed the use of the diversity of higher taxonomic units as indicators of species richness and found significant correlations between the numbers of families and the number of species for certain groups of organisms that were examined over fairly large areas. Beccaloni and Gaston (1995) made a similar comparison between the number of species of a specific butterfly family (Nymphalidae: Ithomiinae) and the total species richness of all other butterflies for Central and South America. Higher-taxon richness as indicator of species richness was found to possess several limitations in tropical areas by Balmford et al (1996). There was little spatial congruence in the distribution and abundance of species of various higher taxa in Britain (Prendergast and Eversham 1997) or in Canada (even negative for mosses and epiphytic lichens, Gould and Walker 1999), while Swiss studies (Obrist and Duelli 1998) found good correlations between species richness of certain taxa as Coleoptera, Diptera and Hymenoptera and total species richness in the samples. When the effort needed for sorting and species determination was included in the latter analysis Heteroptera and vascular plants appeared as most efficient indicators of species richness.

Such endeavours and observations have recently led to a more general theory for the selection of species indicators for more or less distinct communities (Dufrene and Legendre 1997, see also McGeoch and Chown 1998). Algorithms select species that are both highly specific to a site group and widespread within it. The statistical method employed has already found its way into a commercially available software for ordination as ORD (McCune and Mefford 1997). A somewhat related theory relies on the degree of nestedness of more or less fragmented communities (Worthen 1996, Atmar and Patterson 1995 for software); species with high or intermediate level of nestedness may be useful indicators. Some studies show, however, little congruence in nestedness between taxa (e.g. Hansson 1998).

In spite of all these suggestions, there has been little evaluation in the field of how well one or several suggested indicator species do cover the requirements and occurrence of other species. One exception is Nilsson et al. (1995) who showed that the occurrence of the lichen *Lobaria pulmonaria* coincided with occurrences of several other red-listed lichen species. The agreement with the occurrence of red-listed wood beetles was worse but the number of beetle species dependent on hollow trees were larger in sites with *L. pulmonaria*. Abensperg-Traun et al. (1996) found certain potential indicator species to predict very little of the species richness of a West Australian fauna while the inclusion of structural variables as vegetational structural diversity and patch area as covariates considerably improved the predictions.

Functional indicators

As mentioned earlier, indicator systems have been developed for measuring biotic integrity, particularly in limnic ecosystems. The first approach (Karr 1981), based on fish assemblages, considered mainly environmental (water and stream) quality but, as permitting monitoring of ecosystem features, it was also suggested to be useful in surveying the functional biodiversity and thus generally applicable in conservation (Karr 1991). It has been used for practical monitoring during almost a decade in USA. A locally adapted system is being developed to measure the quality and biodiversity of Central American streams (Lyons et al. 1995). As an example, it is not based only on general community composition but on a partitioning of the metrics between various guilds and sensitive species. The authors thus delimit ten measures that are estimated in various streams to indicate biotic integrity. These measures are: Number of native species, percent of benthic species, number of water column species, number of sensitive species, percentage of tolerant species, percentage of exotic species, percentage of omnivores, percentage of native live-bearing species, relative abundance and number of diseased or deformed specimens. Each of these measures is thought to be affected by various types of human disturbance and pollution. A related indicator for areas of high conservation value ("hot spots") only (Winston and Angermeier 1995) is based on the relative densities of the various species (in this case fish) that occurs within a region.

This type of biodiversity analysis may be said to be performed with functional indicators. Such an approach was also suggested by Alard et al. (1994) for grasslands in France. They recognized that grassland vegetation consisted of both indigenous and anthropogenic plant species and that particularly the proportion of competitive species (sensu Grime) indicated changes in general biodiversity. More recently, Angelstam (1998) has proposed a more extensive system of functional indicators for boreal forests. Several plant and animal species, closely dependent on the pristine disturbance regimes of these forests, are supposed to function as indicators of original biodiversity. Similarly, Kuusinen (1996) found that cyanobacterial lichens, including *Lobaria pulmonaria*, indicated old-growth status and long-term continuity of a forest stand. Similar observations were done by Tibell (1992) for crustose lichens in boreal forests. Nilsson and Baranowski (1994) suggested that the number of click beetle species (Elateridae), dependent on hollow trees, were good indicators of mega-tree (and woodland) continuity.

Certain authors have remarked on the great diversity among insects and their potential as an indicator group (e.g. Kremen et al. 1993). As an example, tiger beetles have been suggested as a suitable indicator taxon because of good knowledge of habitat affinities and easy sampling (Pearson and Cassola 1992). Other authors have instead suggested vertebrates and butterflies (and possibly vascular plants) as indicators for gap analyses, i.e. for securing important but underrepresented habitat or ecosystem fragments for conservation by GIS analysis. Again these taxa are assumed to be well-known and to have a precise habitat selection (Scott et al. 1993, however, see Flather et al. 1997 for criticism). Kremen (1992) proposed the use of ordination methods (especially CCA = Canonical Correspondance Analysis) for establishing relationships between the occurrence or abundance of indicator species and environmental factors, especially those related to original and disturbed habitats. Functional indicators can also be more specific: Anderson (1994) suggested the height of a preferred plant species to be used as an indicator of deer browsing pressure and deer effects on plant diversity and, inferentially, also on insect pollinators and herbivores. Deer have also been suggested as suitable indicators of both forest management and landscape quality (Hanley 1996).

Functional indicators may not necessarily be determinants of ecosystem functions even if some authors assume that keystone species would perform particularly well as indicators. In view of present problems with the keystone species concept (e.g. Lawton 1994), less emphasis may be put on such possible relationships. However, functional indicators should be closely related to or strongly dependent on important structures or processes in the ecosystems.

Indicators from hierarchy theory

Noss (1990) advocated an application of hierarchy theory in the selection of indicators. Within each level of organisation, from genetics via species, community, ecosystem, landscape and finally to region, he distinguished three features, composition ("taxonomy"), structure (often equal to spatial distribution) and function (ecological processes). He then observed that hierarchy theory e.g. predicts that higher levels incorporate lower levels and constrain the behaviour of dependent entities. The lower levels contain species identities, abundance and many main functions but higher level properties may emerge, and effects at one level may be expressed in unpredictable ways at another level. One main conclusion was that total biodiversity needs many indicators and several of them may profitably be physical ones, e.g. structures or processes. Indicators may thus be derived from the basic factors or premises for community composition or local biodiversity. More generally, there is a need to monitor indicators of compositional, structural and functional biodiversity at multiple levels of organisation. However, all features and levels cannot be utilised in any realistic system; the most important indicators have to be selected for specific systems and problems. Table 1 lists some potential indicator features for common terrestrial system. They may all be considered as functional indicators.

Focal species

Lambeck (1997) outlined a management approach involving "focal species". He distinguished the (focal) species in a local community, pristine or not, that were most sensitive with regard to 1) area requirement 2) short dispersal distances or connectivity, 3) critical resources (e.g. food or substrate specialisation), and 4) natural or induced processes (e.g. recurrent fires or grazing). If the landscape was managed with regard to structure and function to retain such species then the vast majority of other species should also be thriving. The focal species complement was also reflected in a specific landscape composition, including particular ecosystem processes.

An indicator system based on focal species may be adjusted for pristine landscapes (emphasising the requirements of very specialised species), managed landscapes (considering requirements of the species we want particularly to retain), to "metapopulations" (considering the most sensitive subdivided population) and even to "one-species-systems". Such indicator systems can be applied at various scales, at larger scales probably by stressing connectivity and possibilities for dispersal. Furthermore, if we want it cost-effective we may use only the physical landscape structure as an indicator of what is or will be retained and how to change the landscape in order to get desirable biodiversity.

Table 1. Possible indicator features according to level of organisation and structural (incl. taxonomical or compositional) and functional properties. Original based on hierachy theory, this list could be considered as a "smorgasbord" for functional indicators but actual indicators have to depend on characteristics of the specific system to be monitored. Based on Noss (1990).

Levels	Composition	Structure	Process
Region	Geomorphology Endemism	Heterogeneity Fragmentation	Geomorphic processes Economic processes
Landscape	Patch types β & γ diversity	Connectivity Juxtaposition	Disturbances Movements Patch dynamics
Local ecosystem/community	Species/guilds α-diversity	Biomass Physiognomy	Productivity Herbivory Predation Pollination
Population	Abundance	Dispersion	Natality Mortality Dispersal
Genetic	Allelic diversity	Heterozygosity Effective population size	Inbreeding Drift

Properties of indicator species

The more recent publications have thus often proposed indicator species even if non-living types of indicators have not been completely disregarded (e.g. Faith and Walker 1996). Thus, there are reasons to examine what characteristics are necessary or desirable for species that may serve as (functional) indicators.

Some desirable characteristics, related to general adaptations in ecology and behaviour of the particular species, are: – being specialised on the ecosystem or landscape to be monitored (habitat specialist); – sensitive to artificial disturbance in at least one specific factor, over a wide range of natural variability (reactive); – having fairly large area and resource requirements (spatial coverage); – being fairly common and easily and cheaply identified and sampled (economy).

Some authors suggest that use of indicator species should be independent of sample size or scale (Noss 1990, Weaver 1995) but that is probably too much to hope for.

There are also certain requirements on the spatial distribution of such potential indicator species (cf. Harrison 1991): – they should have continuous and demographically balanced populations, i.e. clumping should not be too severe and particular age classes should not dominate in the system examined; – they should preferably be resident species; – if their populations are characterised by sinks and sources then the monitored habitat should contain source populations; – if they exist as subdivided or fragmented populations then the patches examined should at least at the outset host equilibrium metapopulations.

Similarly, there are requirements on the temporal dynamics. Suitable species should: – consist of populations with rapid density responses to disturbances or habitat changes (i..e. short-lived species) or – monitoring should be focused on reproduction, recruitment or individual health, for plants on growth characteristics (long-lived species) and – the populations examined should not be affected by any conspicuous demographic stochasticity or genetic impoverishment due to long-term marginal population sizes.

It might be noticed that most plant species are long-lived and that short-lived animal species are often naturally characterized by irregular or cyclic fluctuations even in fairly stable environments. A compromise may have to be reached; however, reproduction or physiological condition may often serve as a more reliable indicator than presence or numbers.

Some indicator systems in use

Indicator systems have been developed and used in practical monitoring in limnic environments in USA and other North American countries (Karr 1981) and e.g. in Sweden (Johnson and Wiederholm undated) and have been out-

lined for boreal forests in Canada (McKenney et al. 1994). A system of "signal" species for delimiting old-growth boreal forests was developed from natural history observations in northern Sweden by the local team "Steget före" (Karström 1992) and further analysed by Olsson and Gransberg (1993). It relied on occurrences of certain rare cryptogams and fungi. Preliminary results from an extensive Swedish research project on indicators of forest biodiversity is available in Swedish (Anon. 1999).

Acknowledgements – I appreciate comments by Lena Gustafsson, Gunnar Jansson, Tor-Björn Larsson and Per Sjögren Gulve.

References

Alard, D., Bance, J. F. and Frileux, P. N. 1994. Grassland vegetation as indicator of the main agro-ecological factors in a rural landscape: Consequences for biodiversity and wildlife conservation in central Normandy (France). – J. Environ. Manage. 42: 91–109.

Anderson, R. C. 1994. Height of white-flowered trillium (*Trillium grandiflorum*) as an index of deer browsing intensity. – Ecol. Appl. 4: 104–109.

Anon. 1999. Indikatorer på biologisk mångfald i skogslandskapet. – Skog and Forskning 99: 2: 7–59, in Swedish.

Angelstam, P. 1998. Towards a logic for assessing biodiversity in boreal forests. – In: Bachmann, P., Köhl, M. and Päivinen, R. (eds), Assessment of biodiversity for improved forest planning. Kluwer, pp. 301–313.

Angermeier, P. L. and Karr, J. R. 1994. Biological integrity versus biological diversity as policy directives. – BioScience 44: 690–697.

Abensberg-Traun, M. et al. 1996. Biodiversity indicators in semi-arid, agricultural Western Australia. – Pacific Conserv. Biol. 2: 375–389.

Atmar, W. and Patterson, B. D. 1995. The nestedness temperature calculator; a visual basic program. – AICS Research, Chicago.

Balmford, A., Green, M. J. B. and Murray, M. G. 1996. Using higher-taxon richness as a surrogate for species richness. I. Regional tests. – Proc. R. Soc. Lond. B 263: 1267–1274.

Bader P., Jansson, S. and Jonsson, S. G. 1995. Wood-inhabiting fungi and substratum decline in selectively logged forests. – Biol. Conserv. 72: 355–362.

Beccaloni, G. W. 1995. Predicting the species richness of neotropical forest butterflies: Ithomiinae (Lepidoptera: Nymphalidae) as indicators. – Biol. Conserv. 71: 77–86.

Cairns Jr, J. 1986. The myth of the most sensitive species. – BioScience 36: 670–672.

DeGraaf, R. M. and Chadwick, N. L. 1984. Habitat classification: a comparison using avian species and guilds. – Environ. Manage. 8: 511–518.

Dufrene, M. and Legendre, P. 1997. Species assemblages and indicator species: the need for a flexible asymmetrical approach. – Ecol. Monogr. 67: 345–366.

Faith, D. P. and Walker, P. A. 1996. How do indicator groups provide information about the relative biodiversity of different sets of areas? On hotspots, complementarity and pattern-based approaches. – Biodiv. Lett. 3: 18–25.

Flather, C. H. et al. 1997. Identifying gaps in conservation networks. Of indicators and uncertainty in geographic-based analyses. – Ecol. Appl. 7: 531–542.

Gould, W. A. and Walker, M. D. 1999. Plant communities and landscape diversity along a Canadian arctic river. – J. Veg. Sci. 10: 537–548.

Grime, J. P. 1997. Biodiversity and ecosystem function. The debate deepens. – Science 277: 1260–1261.

Hanley, T. A. 1996. Potential role of deer (Cervidae) as ecological indicators for forest management. – For. Ecol. Manage. 88: 199–204.

Hansson, L. 1998. Nestedness as a conservation tool: plants and birds of oak-hazel woodland in Sweden. – Ecol. Lett. 1: 142–145.

Hansson, L., Fahrig, L. and Merriam, G. (eds) 1995. Mosaic landscapes and ecological processes. – Chapman and Hall.

Harrison, S. 1991. Local extinction in a metapopulation context: an empirical evaluation. – Biol. J. Linn. Soc. 42: 73–88.

Harrison, S. and Fahrig, L. 1995. Landscape pattern and population conservation. – In: Hansson, L., Fahrig, L. and Merriam, G. (eds), Mosaic landscapes and ecological processes. Chapman and Hall, pp. 293–308.

Hawksworth, D. L. and Rose, F. 1976. Lichens as pollution monitors. – Arnold, London.

Järvinen, O. and Väisänen, R. 1979. Changes in bird populations as criteria of environmental change. – Holarct. Ecol. 2: 75–80.

Johnson, R. K. and Wiederholm, T. Undated. State of the environment index – freshwater. – Cent. Environ. Monit., Uppsala.

Jones, S. G., Lawton, J. H. and Schachak, M. 1997. Positive and negative effects of organisms as physical ecosystem engineers. – Ecology 78: 1946–1957.

Karr, J. R. 1981. Assessment of biotic integrity using fish communities. – Fisheries 6: 21–27.

Karr, J. R. 1990. Biological integrity and the goal of environmental legislation: Lessons for conservation biology. – Conserv. Biol. 4: 244–250.

Karström, M. 1992. Steget före – en presentation. – Svensk Bot. Tidskr. 86: 103–114.

Kremen, C. 1992. Assessing the indicator properties of species assemblages for natural areas monitoring. – Ecol. Appl. 2: 203–217.

Kremen, C. et al. 1993. Terrestrial arthropod assemblages: their use in conservation planning. – Conserv. Biol. 7: 796–804.

Kuusinen, M. 1996. Cyanobacterial macrolichens on *Populus tremula* as indicators of forest continuity in Finland. – Biol. Conserv. 75: 43–49.

Lambeck, R. J. 1997 Focal species: a multi-species umbrella for nature conservation. – Conserv. Biol. 11: 849–856.

Landres P. B., Verner, J. and Thomas, J. W. 1988. Ecological use of vertebrate indicator species: a critique. – Conserv. Biol. 2: 316–328.

Lauder, A. E. and Murphy, D. D. 1994. Umbrella species and the conservation of habitat fragments: a case of a threatened butterfly and a vanishing grassland ecosystem. – Biol. Conserv. 69: 145–153.

Lawton, J. H. 1994. What do species do in ecosystems? – Oikos 71: 367–374.

Loiskekoski, M., Halko, P. and Patosaari, P. 1994. Sound forestry – sustainable development. List of criteria and most suitable quantitative indicators. – Ministerial Conference on the protection of forests in Europe, Helsinki.

Lyons, J. et al. 1995. Index of biotic integrity based on fish assemblages for the conservation of streams and rivers in west-central Mexico. – Conserv. Biol. 9: 569–584.

McCune, B. and Mefford, M. 1997. ORD – Multivariate analysis of ecological data. Ver. 3.10. – MJM Software, Gleneden Beach, Oregon.

McGeoch, M. A. and Chown, S. L. 1998. Scaling up the value of bioindicators. – Trends Ecol. Evol. 13: 46–47.

McKenney, D. W. et al. 1994. Towards a set of biodiversity indicators for Canadian forests: proceedings of a forest biodiversity indicators workshop. – Can. For. Serv., Sault St. Marie.

Naeem, S. et al. 1994. Declining biodiversity can alter the performance of ecosystems. – Nature 368: 734–737.

Nilsson, S. G. and Baranowski, R. 1994. Indicators of megatree continuity – Swedish distribution of click beetles (Coleoptera, Elateridae) dependent on holllow trees. – Entomol. Tidskr. 115: 81–97.

Nilsson, S. G. et al. 1995. Tree-dependent lichens and beetles as indicators in conservation forests. – Conserv. Biol. 9: 1208–1215.

Noss, R. F. 1990. Indicators for monitoring biodiversity: a hierarchial approach. – Conserv. Biol. 4: 355–364.

Obrist, M. K. and Duelli, P. 1998. Wanzen und Pflanzen. Auf der Suche nach den besten Korrelaten zur Biodiversitet. – Informationsblatt des Forschungsbereiches Landschaftsökologie 37.

Olsson, G. A. and Gransberg, M. 1993. Indikatorarter för identifiering av naturskogar i Norrbotten. – SNV Report 4276.

Pearson, D. L. and Cassola, F. 1992. World-wide species richness patterns of tiger beetles (Coleoptera: Cicindelidae): indicator taxon for biodiversity and conservation studies. – Conserv. Biol. 6: 376–391.

Prendergast, J. R. and Eversham, B. C. 1997. Species richness covariance in higher taxa: empirical tests of the biodiversity indicator concept. – Ecography 20: 210–216.

Reid, W. V. et al. 1993. Biodiversity indicators for policy-makers. – World Resour. Inst.

Ritchie, M. E. 1999. Biodiversity and reduced extinction risk in spatially isolated rodent populations. – Ecol. Lett. 2: 11–13.

Scott, J. M. et al. 1993. Gap analysis: a geographical approach to protection of biological diversity. – Wildl. Monogr. 123: 1–42.

Severinghaus, W. D. 1981. Guild theory as a mechanism for assessing environmental impact. – Environ. Manage. 5: 187–190.

Skye, E. 1979. Lichens as biological indicators of air pollution. – Annu. Rev. Phytopathol. 17: 325–341.

Tibell, L. 1992. Crustose lichens as indicators of forest continuity in boreal coniferous forests. – Nord. J. Bot. 12: 427–450.

Väisänen, R. and Järvinen, O. 1977. Dynamics of protected bird communities in a Finnish archipelago. – J. Anim. Ecol. 46: 891–908.

Verner, J. 1984. The guild concept applied to management of bird populations. – Environ. Manage. 8: 1–14.

Walker, B. H. 1992. Biodiversity and ecological redundancy. – Conserv. Biol. 6: 18–23.

Weaver, J. C. 1995. Indicator species and scale of observation. – Conserv. Biol. 9: 939–942.

Williams, P. H. and Gaston, K. J. 1994. Measuring more of bio-diversity: can higher-taxon richness predict wholesale species richness? – Biol. Conserv. 67: 211–217.

Wilcove, D. S. 1989. Protecting biodiversity in multiple-use lands: lessons from the US Forest Service. – Trends Ecol. Evol. 4: 385–388.

Winston, M. R. and Angermeier, P. L. 1995. Assessing conserva-tion value using centers of population density. – Conserv. Biol. 9: 1518–1527.

Worthen, W. D. 1996. Community composition and nested-subset analyses: basic descriptors for community ecology. – Oikos 76: 417–426.

Electronic conference on research and biodiversity: preliminary report of the session on forest

Background

Considering that a fluent and direct dialogue between scientists and the wider community is the best way to achieve a consensus, an Electronic Conference on Research and Biodiversity was run from 4 May to 14 june, 1998. Everyone (fishermen, farmers, businessmen, policy-makers, civil servants, land-planners, hunters, students...) and scientists was encouraged to participate.

The Electronic conference was structured according to the six priority areas in the EU Biodiversity Strategy (Communication of the European Commission to the Council and to the Parliament on a European Community Biodiversity Strategy): Conservation of Natural Resources, agriculture, forestry, fisheries, tourism, regional and spatial planning and energy and transport. Each session was chaired by a responsible person who each week encouraged invited experts to comment upon selected topics.

This document is based on a draft text regarding the forest session. See Esteban et al. 1999 for a full report.

Introduction

The forest is the home of a major part of the global biological diversity and the ecological functions of the forests are crucial for our well-being, and e.g. combat climate change and natural disasters. The forest delivers wood and other products to man and is a place to go for the industrial man to enjoy the beauty of nature.

To maintain forest biodiversity – ecosystem processes, functional populations, species diversity and gene pools – we need knowledge. The session on "Forestry" of the e-conference on Research and Biodiversity was designed to give a overview of research needs based on a dialogue on forest biodiversity in a broad European perspective. Special emphasis was to be given on the needs expressed by the forestry sector itself, including relevant governmental and non-governmental organisations. A number of guest contributors, from various fields, were invited in order to present a broad perspective as basis for the discussion. During the conference responses to the invited contributions

were received, but also a number of contributions raising new issues. The geographical width in the participitation was great and this is noteworthy also in the discussion of single topics.

Review of the e-conference session on forests

As the participitation of forestry representatives was quite active in the session on forests it is not surprising – but nevertheless important to have confirmed – that a major part of the contributors stressed the need to develop silvicultural methods that conserve or enhance biodiversity. To avoid repetition we only briefly mention the points raised in the review of contributions below and treat this central theme more in detail under the heading of major research themes in the following section.

Trying to catch the European dimension of forest biodiversity research needs the invited guests of the first round of the forest session presented invited contributors from north, central and south Europe respectively. Börje Pettersson, Chief Ecologist of Stora Forest, opened by sharing some views of a Fennoscandian, export directed, forest company with a relatively long experience of forestry biodiversity issues. He notes that in Swedish forestry considerations to biodiversity are practically implemented in a large scale based on a working hypothesis, developed in dialogue with scientific expertise but that a scientifically based monitoring programme for evaluation is still lacking. The other guest the first week, Erik Sandström of the Swedish governmental body responsible for the national forest policy – also stresses the need of support of biodiversity researchers to implement the national policy. Sandström presents some ideas on how the information flow between scientists and foresters should be organised. Erling Berge from the Norwegian Univ. of Science and Technology responded with some further ideas how the research should be organised and pointed at the need to include political and socio-economic well as the ecological conditions. Lennart Hansson of the Swedish Univ. of Agri-

Tor-Björn Larsson (tor-bjorn.larsson@environ.se), Swedish Environmental Protection Agency, SE-106 48 Stockholm, Sweden.

cultural Sciences stressed the division of responsibilities between academic researchers looking for general mechanisms and theories and the forest manager, who should be educated enough to adapt the principles in his specific forest, of which he has the intimate knowledge. Frederic Gosselin of Cemagref, France, responded to Hansson that room for "applied research" should be given. Gosselin is very skeptical about the ability of local managers to make a sound synthesis between very general ecological theories and local circumstances. He thinks what is lacking is 1) to ask and answer questions in management terms (the "how much is enough" referred to by Suvi Rainio and Jari Niemelä elsewhere in the discussion) and 2) to monitor the manager's options (also much debated in this forum).

The following week three invited contributions from Germany broadened the framework of the discussion. Axel Springer Verlag has played a crucial role as a major customer to put pressure on the producers of pulp and paper to make their production environmentally sound. Florian Nehm, Head of Environmental Management, presented the ecological standards to be met, of which biodiversity is one. In an intervention Simone Matouch, Coordination Group on Biodiversity Research, Austria, added that the certification of forests and wood-products becomes more and more relevant also for European forestry and biodiversity. In the next invited contribution Klaus von Gadow, Professor at the Georg-August-University Göttingen stresses that silviculture at least in central Europe is very intense ("Forest gardening") and that the development of stands can be determined by man, offering a variety of options with respect to biodiversity. This is also reflected in the third invited contribution given by two eminent German forest managers from the Hessische Forstamt Vohl: Eberhard Leicht and Alexander Sinz. They demonstrate clearly their awareness of biodiversity responsibilities according to the Convention on Biological Diversity and Agenda 21 asking for very straightforward knowledge on the effects of silvicultural practices and biotope conservation to develop their forest management optimally.

The journey through Europe was completed the third week by three southern contributions: Kostas A. Spanos, Research scientist at the Forest Research Institute of Thessaloniki opened by stating that conservation of forest biodiversity in the Mediterranean countries has become a very important task in the last decades. Impacts such as overgrazing, fires, urbanization, forest destruction for fuelwood and agriculture land, environmental pollution, heavy tourism load in coastal areas and the islands, global warming and climatic change have strongly affected the total area of forested land. A number of crucial topics require greater understanding and research, to a large extent similar to those raised earlier by the north and central European contributors and/or more specifically identified as a major theme in the following section. Jordi Camprodon, Centre Tecnologic Forestal de Catalunya in a response presented research and conservation of biodiversity in forest

ecosystems of Catalonia aiming at conservation and reversing certain damaging processes in forest habitats. Forest fires, an increasingly hot Mediterranean issue but ecologically important also in north Europe, cf. the section on major research themes below, were commented upon in some depth by Jose M. Moreno, and Federico Fernandez-Gonzalez, from Univ. de Castilla-La-Mancha and Univ. Complutense de Madrid. Orazio Ciancio, Univ. of Florenze in the third south European invited contribution added a new dimension by introducing humanistic aspects. The debate on the values of the forest should be carried on by specialists in different fields: philosophers, theologians, men and women of letters, economists, biologists, ecologists, anthropologists, historians, geographers, environmentalists. Klaus Halbritter, Univ. of Göttingen, in a response to Ciancio states that in approaching the forest as a complex system technocrats seem to have a strong lead. This cannot be completely avoided, e.g. in order to assess natural and human impact on forests, regional, national and supra-national inventories are required.

Next the forest session expanded the geographical perspective with a contribution on tropical forests by Ian Hunter, Director of the European Forest Institute. Hunter starts by stating that there could scarcely be a more stark contrast than the degree of knowledge about and problem with conserving biodiversity than exists between tropical and boreal regions. Biodiversity decreases from south to north from the equator. Knowledge about biodiversity increases from south to north. So poor is the actual knowledge of biodiversity in many tropical rainforest ecosystems that it is not possible to prove, credibly, that any tropical forest harvesting operation is truly sustainable. Unknown aspects of biodiversity may be eliminated in ignorance. Hunter further claims that three crucial areas require greater understanding: 1) a working formula that allows reasonable activity to continue while greater knowledge is sought, 2) continued emphasis on species identification, 3) a major emphasis in understanding the driving mechanisms that determine variety and rarity in tropical forests.

A global perspective of forest biodiversity is further discussed by Tim Boyle working at the Center for International Forestry Research in Indonesia. Boyle points at the widely accepted principle that a protected area system alone cannot adequately conserve the world's biodiversity and that sustainability can only be achieved through a landscape approach to natural resource management. Understanding how biodiversity is maintained at the landscape level is thus critical to managing natural resources sustainably. Nick Ananin from Scotland in a reply partly disagreed with the emphasis on landscapes, having seen examples that the resultant enhancements of biodiversity from such a strategy have been incidental as the real interest was only to improve the public perception of forestry. David Kershaw of Northern Ireland submitted a good compromise by stating that landscape aesthetics – of which many of us have bad experience – need to be integrated, but not mixed up, with biodiversity. Kershaw also claims that it is most useful to consider biodiversity at two scales: site and landscape, and reminding that biodiversity problems are very much scale-related and that research really should recognise this.

Valerie Kapos, World Conservation Monitoring Centre in Cambridge in the third invited contribution on global perspective claims the need of understanding the global distribution of species and ecosystems, their protection status, and the threats to them. An overview of the state of forest systems, their composition and function, is essential for prioritising conservation efforts at the international scale. This is necessary for identifying areas that can be exploited with minimal long term effects on biodiversity. Species richness and ecosystem complexity vary enormously across the globe, management practices must take account of this variation.

A new round in the forest session was started the fifth week by two contributions from Finland, a country of people living close to the forest. As previously the guests represent different sector interests: Suvi Raivio, Biodiversity specialist at the Finnish Forest Industries Federation and Rauno Väisänen, Director of Nature Protection at the Forest and Park Service. The Finnish Forest Industries Federation works to improve the economic prerequisites of its member companies and consider biodiversity aspects important to include in forest and wood-purchasing matters and related environmental and land-use policy, and timber-harvesting and transport policy. Biodiversity information important for forestry relates both to effects of the new forest management techniques and landscape ecological planning. Very specific knowledge is needed, cf. the paragraph on "Development of silvicultural methods" in the following section where the specific points are cited from Suvi Raivio. It is very noteworthy that in a following guest contribution, also presented in the second round of the conference, Stig Ohlsson, Principal Administrative Officer responsible for forest policy at the Swedish Environmental Protection Agency, identifies very similiar and just as specific scientific advice. The previously presented views of the practical German foresters Leicht and Sinz also confirms this demand.

As a response to the demand from practice of precise and quantitative advice Jari Niemelä, of Univ. of Helsinki argues that there is no single answer to such specific questions which are put forward by Raivio (and Ohlsson). Niemelä thinks that we should approach the question of "how much is enough" in steps: 1) examine the ecological requirements of species, 2) set the goals, i.e. decide how many species we want to save, 3) based on this decision leave sufficient material or stands that the desired number of species survives, 4) monitor the success of reaching the goals (species survival), and 5) change procedures, if goals are not met. In this process ecological principles, such as the metapopulation theory, and the precautionary principle need to be considered. And the discussion needs to in-

volve economical aspects of forestry as well.

Rauno Väisänen, considering the biodiversity research needs of National Parks and other protected areas, identifies major gaps of knowledge which are evident as regards the management of the protected area system. This is further discussed as a major research theme below. One of the following contributors, Winfried Bücking, Forstliche Versuchs- und Forschungsanstalt Baden-Württenberg, Germany similarly states that biodiversity is not an absolute but a relative value, that can be evaluated only by comparative studies. Strictly protected or long time unmanaged forests in their characteristic site type, landscape and forest stage patterns may be the standard of potential diversity of the different ecosystems.

In a final contribution to the forest session Eduardo Rojas-Briales, Solsona, Catalonia stresses the need of policy research to find acceptable strategies to manage the biodiversity issues. We should favour a quite integrative picture of the different functions or services that our forests offer to modern society, one of these being biodiversity. Protective measures often forget the economic factor, notably how are policy costs and benefits distributed. Forest policy has been a continuous pillaging of a social minority in the name of general interest.

In addition two major specific points emerged more or less spontaneously in the forest session (they are presented more in detail in the following section on major research themes) here we only acknowledge that: 1) research needs concerning afforestation and deforestation in agricultural landscapes was introduced by Gerard Balent, INRA, France and further discussed by George Hendrey, FACE Program Coordinator and Manfred Klein, Federal Agency for Nature Conservation, Germany. 2) The forest genetical issues were raised by the following south European experts: Phil Aravanopoulos, Aristotle Univ. of Thessaloniki, Kostas A. Spanos, Forest Research Institute of Thessaloniki and Federico Magnani and Gabriele Bucci, Univ. della Basilicata, Potenza and Istituto Miglioramento Genetico Piante Forestali, Firenze. Furthermore, a valuable contribution on forest genetics was received from Anatoly P. Tsarev, Petrozavodsk State Univ., Karelia. In the section on major research themes we for practical reasons mainly cite the contribution of Magnani and Bucci, as it covers the issue in a broad sense. Including genetical aspects adds a dimension to the discussion that only briefly can be presented here.

Finally we acknowledge that methods for monitoring of biodiversity and in particular indicators were put forward as an important research theme by several contributors: Gustaw Matuszewski, Forest Research Institute, Warsaw, Poland; Paul Cannon, CABI Bioscience, U.K.; Jervis Good, Terrascope Environmental Consultancy, Ireland; Hermann Ellenberg, Federal Research Centre for Forestry and Forest Products, Institute for World Forestry, Germany; Klaus Halbritter, Univ. of Göttingen, Germany; Michal Brzeski, Research Institute of Vegetable Crops, Polishing critical thresholds should be a major research area, both generally and with regard to forest ecosystems. Problems connected with evaluation of biodiversity are highly relevant for forests and should be covered both in socio-economic research and in demonstration projects addressing the last two questions.

Conclusions

The contributions to the Forestry session reveal a gold-mine of ideas and experiences that should be more fully explored in future activities and continued dialogue. If we return to the EU Biodiversity Strategy and its presentation of the policy area "Forests" it is striking how well the objectives for action (sustainable management of forests, reforestation strategy, adaption to climate change and developing monitoring techniques) are reflected in the contributions in the forest session.

The outcome of the Forestry session supplements the research agenda "Understanding Biodiversity" presented by the European Working Group on Research and Biodiversity EWGRB by reflecting the sector needs. A number of general biodiversity issues, like evaluation of biodiversity, assessment and indicator development were raised in the e-conference but are treated more in detail in the EW-GRB research agenda.

The forests session of the e-conference on Biodiversity clearly demonstrates a demand from managers and policymakers of scientific support and a scientific potential to perform this research.

Acknowledgements – The project was financed by the Autonomous Government of Catalonia, and the European Commission (DG XII). The design and organization of this activity is done by the Ministry of Environment of the Autonomous Government of Catalonia (Conference manager Aniol Esteban), together with the European Working Group on Research and Biodiversity EWGRB. Six members of the EWGRB chaired the conference sessions; 1) Agriculture: Allan D. Watt, Institute of Terrestrial Ecology, Edinburgh, U.K. 2) Forestry: Tor-Björn Larsson, Swedish Environmental Protection Agency, Stockholm, Sweden. 3) Fisheries: Mark Costello, Ecological Consultancy Services Ltd (Ecoserve), Dublin, Ireland. 4) Conservation of natural resources: Andreas Troumbis, Univ. of the Aegean, Lesbos Island, Greece. 5) Tourism and hunting: Linus Svensson, Lund Univ., Sweden. 6) Regional and spatial planning. Energy and transport: Peter Nowicki, European Center for Nature Conservation, Tilburg, Netherlands.

References

Esteban, J. A. et al. 1999. Research and biodiversity. A step forward. Report of an electronic conference. – Ministry of Environment, Goverment of Catalonia, Barcelona.

ECOLOGICAL BULLETINS

ECOLOGICAL BULLETINS are published in cooperation with the ecological journals Ecography and Oikos. Ecological Bulletins consists of monographs, reports and symposia proceedings on topics of international interest, often with an applied aspect, published on a non-profit making basis. Orders for volumes should be placed with the publisher. Discounts are available for standing orders.

ECOLOGICAL BULLETINS still available.

Prices excl. VAT and postage.

41. *The cultural landscape during 6000 years in southern Sweden - the Ystad project* (1991). Editor B. E. Berglund. Price £ 21.00.

42. *Trace gas exchange in a global perspective* (1992). Editors D. S. Ojima and B. H. Svensson. Price £ 5.00.

43. *Environmental constraints of the structure and productivity of pine forst ecosystems: a comparative analysis* (1994). Editors H. L. Goltz, S. Linder and R. E. McMurtie. Price £ 5.00.

44. *Effects of acid deposition and tropospheric ozone on forest ecosystems in Sweden* (1995). Editors H. Staaf and G. Tyler. Price £ 5.00.

45. *Plant ecology in the subarctic Swedish Lapland* (1996). Editors P. S. Karlsson and T. V. Callaghan. Price £ 5.00.

48. *The use of population viability analyses in conservation planning* (2000). Editors P. Sjögren-Gulve and T. Ebenhard. Price £ 25.00.

49. *Ecology of woody debris in boreal forests* (2001). Editors B. G. Jonsson and N. Kruys. In press.

50. *Biodiversity Evaluation Tools for European forests* (2001). Coordinator T.-B. Larsson. Price £ 35.00.